A-LEVEL
AND AS-LEVEL

LONGMAN
REVISE
GUIDES

BIOLOGY

Alan Cornwell
Ruth Miller

LONGMAN A-LEVEL AND AS-LEVEL REVISE GUIDES
Series editors
Geoff Black and Stuart Wall

Titles available:
Accounting
Art and Design
Biology
Business Studies
Chemistry
Computer Studies
Economics
English
French
General Studies
Geography
German
Mathematics
Modern History
Physics
Psychology
Sociology

Longman Group Limited
Longman House, Burnt Mill, Harlow,
Essex CM20 2JE, England
and Associated Companies throughout the world.

© Longman Group UK Limited 1990
Second edition © Longman Group Limited 1995
All rights reserved; no part of this publication
may be reproduced, stored in a retrieval system,
or transmitted in any form or by any means, electronic,
mechanical, photocopying, recording, or otherwise
without either the prior written permission of the
publishers or a licence permitting restricted copying in
the United Kingdom issued by the Copyright Licensing
Agency Ltd, 90 Tottenham Court Road,
London, W1P 9HE.

First published 1990
Second edition 1995

British Library Cataloguing in Publication Data
A catalogue record for this book is available from the British Library.

ISBN 0 582 27771 X

Set by 26 in 10/12pt Century Old Style

Produced by Longman Singapore Publishers (Pte) Ltd
Printed in Singapore

CONTENTS

	Editors' preface	iv
	Acknowledgements	iv
	Names and addresses of the exam boards	v
1	Examination topics and courses	1
2	Examination and assessment techniques	5
3	The diversity of organisms	11
4	Cell biology	25
5	Autotrophic nutrition	41
6	Heterotrophic nutrition	55
7	Homeostasis	69
8	Respiration and gas exchange	85
9	Chemical co-ordination in plants	101
10	Co-ordination in animals	113
11	Transport in animals and plants	131
12	Reproduction and development in mammals	153
13	Reproduction and development in flowering plants	167
14	Support, movement and locomotion	183
15	Genetics and evolution	201
16	Ecology	229
17	Biotechnology and genetic engineering	249
18	Microbiology	263
	Index	277

EDITORS' PREFACE

Longman A-level Revise Guides, written by experienced examiners and teachers, aim to give you the best possible foundation for success in your course. Each book in the series encourages thorough study and a full understanding of the concepts involved, and is designed as a subject companion and study aid to be used throughout the course.

Many candidates at A-level fail to achieve the grades which their ability deserves, owing to such problems as the lack of a structured revision strategy, or unsound examination techniques. This series aims to remedy such deficiencies, by encouraging a realistic and disciplined approach in preparing for and taking exams.

The largely self-contained nature of the chapters gives the book a flexibility which you can use to your advantage. After starting with the background to the A-level, AS-level and Scottish Higher courses and details of the syllabus coverage, you can read all other chapters selectively, in any order appropriate to the stage you have reached in your course.

Geoff Black and Stuart Wall

ACKNOWLEDGEMENTS

We are indebted to the following examination boards for permission to reproduce questions which have appeared in their examination papers.

The Associated Examining Board; Northern Ireland Council for Curriculum, Examinations and Assessment; Oxford and Cambridge Schools Examination Board; University of London Examinations and Asessment Council; University of Oxford Delegacy of Local Examinations; Northern Examinations and Assessment Board.

The examination boards are not responsible for the suggested answers to the questions. Full responsibility for these is accepted by the authors.

The authors would like to thank Vicky Cornwell for invaluable assistance in the preparation of the manuscript and Joanna Becker for many of the student's answers. We would also like to thank Eric Turner for his expert advice.

NAMES AND ADDRESSES OF THE EXAM BOARDS

Associated Examining Board (AEB)
Stag Hill House
Guildford
Surrey GU2 5XJ

University of Cambridge Local Examinations Syndicate (UCLES)
Syndicate Buildings
1 Hills Road
Cambridge CB1 2EU

Northern Examinations and Assessment Board (NEAB)
Devas St
Manchester M15 6EX

University of London Examinations and Assessment Council (ULEAC)
Stewart House
32 Russell Square
London WC1B 5DN

Northern Ireland Council for Curriculum, Examinations and Assessment (NICCEA)
Beechill House
42 Beechill Road
Belfast BT8 4RS

Oxford and Cambridge Schools Examination Board (OCSEB)
Purbeck House
Purbeck Road
Cambridge CB2 2PU

Oxford Delegacy of Local Examinations (ODLE)
Ewert Place
Summertown
Oxford OX2 7BZ

Scottish Examination Board (SEB)
Ironmills Road
Dalkeith
Midlothian EH22 1LE

Welsh Joint Education Committee (WJEC)
245 Western Avenue
Cardiff CF5 2YX

CHAPTER 1
EXAMINATION TOPICS AND COURSES

MODULAR SYLLABUSES

A-LEVELS AND SCOTTISH HIGHERS

AS-LEVELS

TYPES OF ASSESSMENT

GETTING STARTED

Advanced-level Biology syllabuses contain a core of topics which are common to them all, so the variations which occur are minor ones, involving differences in emphasis or in the scheme of examination. The syllabus sections may vary in the way they are set out, in their titles and in how the different topics are combined. The chapters in this book cover the content which is common to *all* the syllabuses and also a selection of topics which are included in most of them. This content will also be found in the other Biology syllabuses which are of equivalent standard, such as the Business/Technical Education Council (BTEC) and the Scottish Higher examinations.

Most candidates for A-level Biology now have a more general but less detailed scientific background than previously. A-level syllabuses have been modified to take this into account. All the current syllabuses incorporate the agreed common core for A-level Biology but, overall, the syllabus content has been reduced. The other major changes are in the schemes of assessment where there is less emphasis on essay style questions.

At GCSE, 20 per cent of the marks are awarded for teacher-assessed practical work and this emphasis on the assessment of coursework is carried through to A-level. Many candidates will welcome the chance to gain credit for good work that they do throughout the course.

ESSENTIAL PRINCIPLES

MODULAR SYLLABUSES

Most of the A-level Biology syllabuses for examination in June 1996 are modular. Some modules are already being set and the full range of modules will be examined in June 1996 and at each examination thereafter. With the modular examination there is some element of choice as to which modules are studied. Some modules are, however, compulsory as the set of modules taken by any candidate must cover the common core. Modules can be banked for up to four years and any module can be re-taken to improve a candidate's grade. All the modular courses include a synoptic element where knowledge and skills from different modules are brought together. It should be noted that although the organisation of content and the method of examination is different for a modular A-level, the range and depth of knowledge and skills has not changed. The major topics listed in Table 1.1 still form the majority of A-level syllabuses.

> The emphasis is on the use and application of biological principles

A-LEVELS AND SCOTTISH HIGHERS

Table 1.1 gives a broad indication of the major topics covered by the various A-level and Scottish Higher syllabuses; it also indicates the chapters in this book which are relevant to each syllabus. It should be noted that titles of syllabus sections do vary from syllabus to syllabus: for example, 'Diversity of Organisms' is sometimes called 'Variety of Life'.

CHAPTER AND TOPIC	Associated Examining Board (AEB)	Cambridge Syndicate (UCLES)	Northern Board (NEAB)	London Board (ULEAC)	Northern Ireland (NICCEA)	Oxford Delegacy (ODLE)	Oxford and Cambridge (OCSEB)	Southern Universities (JBSE)	Welsh Joint Examinations (WJEC)	Scottish Higher (SEB)
3 DIVERSITY OF ORGANISMS/VARIETY OF LIFE	✓	✓	✓	✓	✓	✓	✓	✓	✓	
4 CELL BIOLOGY/CELLS AND ORGANELLES	✓	✓	✓	✓	✓	✓	✓	✓	✓	✓
5 AUTOTROPHIC NUTRITION/PHOTOSYNTHESIS	✓	✓	✓	✓	✓	✓	✓	✓	✓	✓
6 HETEROTROPHIC NUTRITION	✓	✓	✓	✓	✓	✓	✓	✓	✓	
7 HOMEOSTASIS	✓	✓	✓	✓	✓	✓	✓	✓	✓	✓
8 RESPIRATION AND GAS EXCHANGE/RELEASE OF ENERGY	✓	✓	✓	✓	✓	✓	✓	✓	✓	✓
9 CHEMICAL CO-ORDINATION IN PLANTS	✓	✓	✓	✓	✓	✓	✓	✓	✓	✓
10 CO-ORDINATION IN ANIMALS	✓	✓	✓	✓	✓	✓	✓	✓	✓	✓
11 TRANSPORT IN ANIMALS AND PLANTS	✓	✓	✓	✓	✓	✓	✓	✓	✓	✓
12 REPRODUCTION AND DEVELOPMENT IN MAMMALS	✓	✓	✓	✓	✓	✓	✓	✓	✓	✓
13 REPRODUCTION AND DEVELOPMENT IN FLOWERING PLANTS	✓	✓	✓	✓	✓	✓	✓	✓	✓	✓
14 SUPPORT, MOVEMENT AND LOCOMOTION	✓	✓	✓	✓	✓	✓	✓	✓	✓	✓
15 GENETICS AND EVOLUTION	✓	✓	✓	✓	✓	✓	✓	✓	✓	✓
16 ECOLOGY	✓	✓	✓	✓	✓	✓	✓	✓	✓	✓
17 BIOTECHNOLOGY AND GENETIC ENGINEERING	✓	✓	✓	✓	✓	✓	✓	✓	✓	
18 MICROBIOLOGY		✓	✓	✓		✓		✓	✓	

Table 1.1 Syllabus coverage chart: A-level and Scottish Higher examinations

> Table 1.1 outlines the topics relevant for A-level and Scottish Higher.

AS-LEVELS

In 1989 advanced supplementary-level examinations were sat for the first time. AS-level syllabuses are shorter than A-level ones; they either cover fewer syllabus areas, or there is a reduction in overall content (see Table 1.2). It should be emphasised that the topics are examined at A-level depth, and we have included a number of AS-level questions in this book.

CHAPTER AND TOPIC	AEB	UCLES	NEAB	ULEAC	ODLE	OCSEB	JBSE
3 DIVERSITY OF ORGANISMS/VARIETY OF LIFE		✓		✓		✓	✓
4 CELL BIOLOGY/CELLS AND ORGANELLES	✓	✓	✓	✓	✓	✓	✓
5 AUTOTROPHIC NUTRITION/PHOTOSYNTHESIS	✓	✓	✓	✓	✓	✓	✓
6 HETEROTROPHIC NUTRITION	✓				✓		
7 HOMEOSTASIS					✓	✓	
8 RESPIRATION AND GAS EXCHANGE/RELEASE OF ENERGY	✓	✓	✓		✓	✓	✓
9 CHEMICAL CO-ORDINATION IN PLANTS		✓		✓	✓	✓	✓
10 CO-ORDINATION IN ANIMALS	✓	✓		✓	✓	✓	✓
11 TRANSPORT IN ANIMALS AND PLANTS	✓	✓		✓		✓	
12 REPRODUCTION AND DEVELOPMENT IN MAMMALS				✓			
13 REPRODUCTION AND DEVELOPMENT IN FLOWERING PLANTS							
14 SUPPORT, MOVEMENT AND LOCOMOTION							
15 GENETICS AND EVOLUTION		✓	✓	✓	✓	✓	✓
16 ECOLOGY		✓		✓		✓	
17 BIOTECHNOLOGY AND GENETIC ENGINEERING	✓	✓	✓		✓		✓
18 MICROBIOLOGY	✓	✓		✓	✓		✓

Table 1.2 Syllabus coverage chart: AS-levels

TYPES OF ASSESSMENT

All examination boards set papers with structured questions and essays, although the trend is to reduce the number of open-ended essays that candidates are required to attempt. Many boards also set questions which require short answers. These papers are printed in books so that the answers can be written on the question paper, where printed lines or spaces are provided for the answers.

Most examination boards have a scheme of assessment for the practical work done during the course, and whilst some boards still set a practical test, it is unlikely that these will continue when the syllabuses are revised.

Table 1.3 gives a quick reference to the different styles of papers and types of assessment that are used by different boards in A-level and Scottish Higher examinations.

> Table 1.2 outlines the topics relevant for AS-level

	AEB	UCLES	NEAB	ULEAC	NICCEA	ODLE	OCSEB	JBSE	WJEC	SEB
Structured questions/Short answers	✓	✓	✓	✓	✓	✓	✓	✓	✓	✓
Essays/Free response	✓	✓	✓	✓	✓	✓	✓		✓	✓
Practical examination		1✓	2✓	✓		✓	✓		✓	
Teacher-assessed coursework	✓	1✓	✓	✓			✓			✓
Multiple choice		✓	✓		✓					✓
Compulsory project (practical investigation)		✓				✓				
Investigative assignment		✓					✓			

Table 1.3 Types of assessment: A-level and Scottish Higher examinations

1 Alternatives; candidates take a practical examination or teacher-assessed coursework is submitted.
2 For external/part-time candidates; internal candidates submit teacher-assessed coursework.

Note: AEB and London have a written alternative to teacher-assessed coursework. Oxford and Cambridge have optional personal work as part of the assessment.

Table 1.4 gives the same information for AS-level syllabuses.

	AEB	UCLES	NEAB	ULEAC	ODLE	OCSEB	JBSE
Structured questions/Short answers	✓	✓	✓	✓	✓	✓	✓
Essays/Free response	✓	✓	✓	✓	✓	✓	✓
Teacher-assessed practical coursework	✓	✓	✓	✓	✓	✓	✓
Investigative assignment			✓			✓	✓
Multiple choice			✓				

Table 1.4 Types of assessment: AS-levels

See footnote to Table 1.3

Note: AS students with London and Cambridge take papers which are common to both AS and A-level. The Oxford and Cambridge and Cambridge syllabuses include options.

CHAPTER 2

EXAMINATION AND ASSESSMENT TECHNIQUES

EXAMINATION PREPARATION

EXAMINATION STRATEGY

TEACHER ASSESSMENT

PROJECTS

GETTING STARTED

All biology examinations at A-level or its equivalent (Advanced Supplementary, BTEC, Scottish Higher, etc.) involve papers of various types. **Essay** papers (or sections of papers) are set to test your ability to develop arguments or to give expanded detail of topics. **Structured short-answer papers** test the breadth of your knowledge, as do **multiple-choice** questions or papers. Practical work is usually tested by **teacher assessment**, sometimes along with a written paper testing your knowledge of practical techniques. At present some examination boards still set a practical examination but, as the new syllabuses come out, these are disappearing. Some syllabuses also include a **project** or long-term investigation; this can be optional. The particular combination of papers set will vary from syllabus to syllabus, as will the form of practical assessment. You should check with your own syllabus for exact details of the assessment you will be following. Throughout your period of working towards the examination you should check your **understanding**. Your time may not be well spent rewriting notes you have taken during lessons, but you should read through the notes as soon as possible after the lesson is finished to make sure that you understand them. If there are points that do not make sense, refer to the textbooks and/or go and see your tutor. Having made sure that you do understand the current work, you will have a firm basis for understanding future work. You will also have completed the first stage in remembering the material over the longer term.

CHAPTER 2 EXAMINATION AND ASSESSMENT TECHNIQUES

ESSENTIAL PRINCIPLES

EXAMINATION PREPARATION

"Consistent work is the basis for success"

The next step is to **revise regularly**. If you do not look at previously understood material for several months, you will have forgotten much of the detail and, probably, much of your understanding of it. **Repetition** is an important tool in learning and in preparing for examinations. Revision is the essence. It is foolish to leave such learning until the last few weeks. Success comes through **consistent** work rather than an attempt to commit everything to memory in, and for, a short time.

KEYWORDS

One technique which many people find useful is to make cards with keywords, diagrams and brief notes. These can be written soon after a lesson and stored as a file. They can be referred to easily and can be used as 'memory joggers'.

PAST QUESTIONS

Another technique which is very valuable is to make use of past questions set by the examination board whose examinations you are taking. Your school or college will have past papers, or if you are an independent student you can obtain papers by writing to the Publications Department of the relevant examination board.

After studying each syllabus section, you will find it useful to attempt questions based on the material. The most valuable way of doing this is to revise the section and then to attempt an answer to questions **without reference to your notes**. Occasionally, write a **full answer** to the question, keeping to the time allowance you would have in the examination; for other questions, just **plan** answers. In each case, when you have completed the answer or the plan, **check with your notes** to make sure that you have remembered and used facts correctly and have not omitted any relevant information. This part of the process is very important because to improve your understanding you must learn from mistakes, whether they be errors or omissions.

"Learn from any mistakes"

The answers given to questions in this book are not model answers but illustrate one way of selecting and organising material. If your answers differ from the ones given, compare the two and decide whether or not your answer is along the right lines.

EXAMINATION STRATEGY

"Follow all instructions"

When taking examinations, the first essential when you receive the question paper is to **read the instructions carefully**; they must be followed to the letter. Check how many questions have to be answered; not only the total number but also how many from each section of the paper. Also check whether or not there are any compulsory questions. In **short-answer papers**, the questions will vary in length, so there is no point in dividing the total time available by the number of questions. However, this division of time is very important in **essay** papers. In any examination, be sure to answer the correct number of questions. Remember that if you answer only four questions when five should be answered, your maximum mark is 80 per cent not 100 per cent, so you have diminished your chances of doing well. The effect is even greater if you answer three instead of four questions. The rest of this section gives guidance on how to organise your time and effort in the different types of examination papers.

ESSAYS

Choose your questions

Take care in choosing the questions that you intend to answer. Read through the whole paper and mark the questions that you think you can answer. Then, read through your chosen list again and pick the appropriate number of questions (taking into account any instructions about the number of questions per section).

CHAPTER 2 ESSENTIAL PRINCIPLES

Plan your answers

> A plan can help

When you start to answer each question, plan it out first and write down your plan. This will not only help you to organise your answer logically but will also give you a checklist to which you can refer whilst answering. In this way you will be less likely to repeat yourself or to miss out important sections. When questions are subdivided, use the *mark allocations* given on the exam paper as a guide to the detail needed for each section. In open essays, you can place your own emphasis in your answer, but ensure that you cover all the aspects that you have planned.

Write clearly

Write your answers clearly and carefully. State your facts and express your ideas in detail. Whenever possible, qualify your statements with further details or examples to show that you really understand what you are writing. Do not give the impression that you have just remembered a few 'jargon' words. If the question is divided into parts a), b)i), b)ii), etc., your answer must be similarly divided.

Answer relevantly

> Answer the question actually set

Make sure that you keep your answer relevant. Good biology will not obtain any marks if it is not relevant to the question set. Frequent reference to your plan should help you to keep to the question and not be sidetracked into other topics.

Organise your time

Organise your time so that you are able to read through your answers when you have completed the paper to check for errors and omissions.

SHORT-ANSWER PAPERS

Check question styles

These are set in a variety of styles so before sitting the examination use past papers to familiarise yourself with the types of questions set by your examination board.

Keep to time

Assess the relationship between minutes and marks to get an indication of how long you can afford to spend on each question. Work your way through the paper, but if you come across a question which, after reading it a couple of times, you feel you cannot answer, pass on and come back to it later.

Answer appropriately

Remember that the **leader lines** or **spaces** left on the question paper indicate the length of the answer which is expected. Read the questions carefully and do what is asked. If a question says 'list ...' do not waste time **describing**; if a question gives two lists of data and asks you to graph **one** list do not graph **both**, and so on. If you are asked to calculate, show your workings, as most of the marks will be for your method and not just for the answer. As with essay papers, write clearly and concisely, trying to answer all the questions in the time allowed.

MULTIPLE CHOICE

Work quickly

All the questions will be compulsory and you will have to work quickly to complete this paper. Some questions will be quite straightforward but others will be more complex. With these, make notes or draw diagrams rather than trying to work them out in your head.

Read the stem carefully

> Watch out for a negative stem

Always read the stem carefully and make sure that you do not miss vital words. Sometimes questions have a **negative** stem and it is particularly important to notice this. The format would be 'Which of the following does *not* ...'. When you have chosen a response, have another look at the stem before you put your response on the computer sheet.

CHAPTER 2 EXAMINATION AND ASSESSMENT TECHNIQUES

Keep answers in line
In this type of paper there will inevitably be some questions which you will find easier than others. If you find a question which after a minute or so is still unanswerable, leave it and go on to the next. Remember to mark the question you have missed out and be careful not to get your answers out of line on the computer sheet.

'Educated' guesses
When you go back to a question, if you still cannot decide on the correct response, *eliminate* any that you are sure are wrong and then make an 'educated' guess from those still remaining. Should the questions have two or three alternatives remaining, a guess would have a 50 per cent or $33\frac{1}{3}$ per cent chance of being correct, and so would be better than a blank!

PRACTICAL PAPERS

Check for special instructions
When you are given the paper, read through the instructions and make a special note of whether they suggest starting any particular question at the beginning of the examination. If they do, then follow the instruction as it means that the question will take a long time, often with periods when you are waiting for enzymes to work or tissues and solutions to equilibrate. These are periods during which you can get on with other questions while you are waiting.

Do as question asks
Read all the questions carefully and make sure you understand what you have to do before you start. Take particular note of the wording of questions and do what they say. For example:

- if you are asked to **tabulate**, the information should be given in a table, *not* as a series of sentences
- if you are asked to **annotate a drawing**, the comments should be alongside the labels, *not* as a separate statement
- if you are asked to **make a key** to separate a group of specimens, this means using dichotomous separations, *not* just writing a table of differences

Drawings
Drawings should be large and clear. Label lines should be drawn with a rule and should not criss-cross. Do not put arrowheads on label lines. Labels should be neat and ideally should be placed under each other. When drawing plans of microscope sections make sure that the tissues are shown as clearly marked areas (not collections of cells) and that they are correctly proportioned. High-power microscope drawings of cells should be made with the cell components correctly proportioned and the different types of cells correctly sized relative to each other.

Complete the paper
As with all the other types of examinations it is important that you complete the paper. The questions may not each carry the same number of marks and so again use the mark allocations as a guide to the relative amount of time you should spend on each question. With practical examinations however, the amount of time you need to spend on a question will also depend on your own practical skills. Some techniques will be easier for you and so you will be able to allow more time for questions which do not involve your own particular strengths.

TEACHER ASSESSMENT

The practical element of the assessment will be partly or wholly by teacher-assessed work during the course of your studies. Some boards set specific pieces of practical work for assessment, others have the assessment based on the on-going practical work associated with the syllabus. You should check the requirement of your particular board.

CHAPTER 2 ESSENTIAL PRINCIPLES

Whatever method is used, the basic essential is the development of an **efficient technique** for practical work and of the **basic skills** needed for practical biology, all of which will be assessed. You will need to show your ability to:

> **Skills to be tested**

- handle apparatus and materials
- observe and measure accurately
- formulate hypotheses
- design and carry out investigations
- handle data and carry out statistical analyses
- record and communicate results and conclusions

The following are examples of areas which may be tested in teacher assessment:

MICROSCOPY

You will need to be able to set up and use a microscope at a range of magnifications and to record your observations by drawing and possibly by using a micrometer eyepiece to make measurements. You may be expected to make temporary preparations and use appropriate staining techniques.

EXPERIMENTAL LABORATORY WORK

A whole range of experimental work, covering biochemistry as well as both plant and animal physiology, will be included. In this work you will need to show your ability to carry out a suitable method and to record it in detail. You will have to record your results and draw valid conclusions from them. Your discussion of results should include consideration of the biological principles involved and their significance in the life of organisms.

FIELDWORK

You will need to show your ability to select and use suitable methods for obtaining numerical as well as qualitative data. You will have to use and perhaps produce keys for identifying both animals and plants. Results should be presented in both tabular and graphical form and you should subject your data to simple statistical treatment. Your performance will be judged on the appropriateness of your techniques, the suitability of your chosen form of data presentation and your discussion of the significance of your findings.

PROJECTS

Many A-level syllabuses include either a compulsory or an optional project as part of the assessment. If your examination includes a project the syllabus will contain all the necessary details and it is wise to study these before starting work on it. Usually a suggested format for the report is given and also a list of criteria against which the project will be marked. You will find these very useful as they indicate what information and discussion you should include. The normal procedure for choice of topic is for you, in consultation with your tutor, to choose an area of investigation which interests you. This, usually with a brief synopsis, will be submitted to the examination board for approval. When choosing a topic, it is very important to bear in mind the amount of time you will have available and select your topic accordingly. Remember that carrying out experiments or collecting field data may often take longer than you anticipate.

Writing up your project

When writing up your project be sure that you know the length that is expected and bear this in mind all the time. The assessment will be based mainly on your **own work** and your **own analysis** of it. So, although you will have to research **published reports** covering your field of study, you should mention these only briefly when writing up your project. The emphasis should be on your own methods, statistical analysis of your data and any relevant discussion.

POSTSCRIPT

From the beginning of your course, **aim high**. Make the assumption that all the topics are important. Some you will find more difficult than others, but persevere with them; do not be discouraged. Nobody finds the whole of a subject easy and there is a great feeling of achievement when you get on top of an area of difficulty. Looking back on it you will have a sense of satisfaction and will probably wonder why, at one time, you found it difficult.

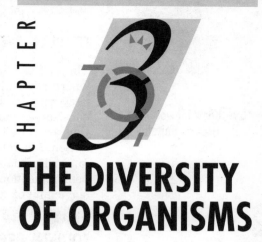

CHAPTER 3
THE DIVERSITY OF ORGANISMS

LEVELS OF ORGANISATION

PRINCIPLES OF CLASSIFICATION

THE FIVE KINGDOMS

PRACTICAL WORK

GETTING STARTED

This chapter deals with **levels of organisation**, the concept of the **species** and the principles of **classification** before considering a range of different organisms. It is essential to check the relevant syllabus very carefully so that you are quite clear what is required. Some syllabuses are very specific about the organisms included for study and types are listed, but others refer to groups of organisms only. It is also worth checking all sections of the syllabus for reference to specific organisms, not just a section headed 'Variety of Life' or 'Diversity of Organisms'. In some syllabuses, the diversity of organisms is linked to the fieldwork.

Most syllabuses give some direction as to how the organisms should be studied; the emphasis is usually on the recognition of external features characteristic of the group to which they belong. An understanding of the different levels of organisation is necessary in order to appreciate the taxonomic position of species, and also to be capable of identifying specimens using keys or guides. The external features of an organism can also be used to show how it is adapted to its environment and to its mode of life. It should be appreciated that in order to gain a good understanding of this section of the syllabus, reference should be made to actual specimens wherever possible. A great deal of benefit can be derived from making drawings of living or preserved specimens with clear labels and annotations of the features displayed.

CHAPTER 3 THE DIVERSITY OF ORGANISMS

ESSENTIAL PRINCIPLES

LEVELS OF ORGANISATION

Know the various levels of organisation

CELLULAR LEVEL

The **cellular** level of organisation is represented by unicellular plants and animals, where the entire body consists of a single cell able to carry out all the functions necessary for life. The term 'cell' is used for the basic unit of an organism and it consists of **cytoplasm** containing a **nucleus**, together with a number of **organelles**. Cell structure is dealt with in more detail in Chapter 4.

Some members of the protozoa, which are all **unicellular**, have a very complex internal organisation, with organelles specialised for the ingestion of food, digestion, osmoregulation and locomotion. These organisms show division of labour within one single unit of cytoplasm.

COLONIAL ORGANISATION

Several groups of organisms show colonial organisation, involving **aggregations of unicells**. This is seen particularly clearly in the green alga *Volvox*, where the unicells are connected by cytoplasmic strands and some of the cells in the colony have a reproductive function. A similar organisation is seen in the sponges, to which reference will be made later in the chapter.

TISSUE LEVEL

Some **multicellular** organisms illustrate the tissue level of organisation; this is seen most clearly in the Cnidarians, where there are two body layers, the **ectoderm** on the outside and the **endoderm** on the inside. Most of the cells making up the ectoderm are similar, forming an external layer, with some cells specialised as stinging cells and sensory cells. Similarly, the endoderm consists of cells specialised for the digestion and absorption of food. There is no organisation of the tissues into organs specialised for a definite function.

ORGAN DEVELOPMENT

The development of **organs** allows certain parts of an organism to carry out specific functions, and results in a much greater division of labour. It is associated with the origin of a third layer in the body of the embryo, the **mesoderm**, which is characteristic of **triploblastic** animals. The mesoderm has become the layer which makes up most of the body and gives rise to many important organ systems. This level of organisation results in an increase in both the size and complexity of organisms.

PRINCIPLES OF CLASSIFICATION

Important groups used in classification

CONCEPT OF THE SPECIES

A **species** is a group of organisms which share a large number of common characteristics and which can breed amongst themselves but not usually with any other species. Sometimes species are further divided into **subspecies** or **varieties**, which are capable of interbreeding but are separated from each other by structural differences.

Similar species are grouped together into a **genus**, similar genera into **families**, families into **orders**, orders into **classes** and classes into **phyla**. Phyla are subdivisions of a **kingdom**. All the organisms in a kingdom have basic features in common. Until fairly recently it was common to divide all living organisms into two kingdoms, the **plant kingdom** and the **animal kingdom**, but there are some organisms which can be classified in both kingdoms and some which do not fit into either. A number of alternative schemes have been suggested and the one which appears to have most support at the moment is a modification of Whittaker's Five Kingdom system by Margulis and Schwartz. The five kingdoms are:

CHAPTER 3 ESSENTIAL PRINCIPLES

> The five kingdoms

- Prokaryotae
- Protoctista
- Fungi
- Plantae
- Animalia

BINOMIAL SYSTEM

Organisms are named according to the **Binomial System** introduced by **Linnaeus** in 1753. Each is named by its genus and species, the generic name, in Latin, being written first, in italics with a capital letter, followed by the specific name, also in Latin and italics, but with a lower-case initial letter. The earthworm belongs to the genus *Lumbricus* and the species is *terrestris*, so its name, written in full, is *Lumbricus terrestris*. The scientific name of an organism enables precise identification worldwide, whereas the common name does not. Many flowering plants have different common names in different parts of the country.

KEYS

Organisms are identified using keys, which are based on **differences** between them. The use of keys is a valuable exercise for a student and in some of the schemes of assessed practical work it is a requirement. It is equally valuable to be able to construct keys to separate organisms, and this type of exercise is frequently set in practical tests. The candidates are given a range of organisms, asked to select visible features which are different and then required to construct a key using those features. In making a key it is important to choose easily observable, clear-cut differences between organisms, and to avoid variable features such as size or colour.

THE FIVE KINGDOMS

KINGDOM PROKARYOTAE

The members of this kingdom are **bacteria**, either non-photosynthetic such as *Escherichia coli*, or photosynthetic, such as *Nostoc*. These organisms have no internal cell membranes; there is no nuclear membrane, no endoplasmic reticulum, no mitochondria and no Golgi body. Most A-level syllabuses do not require a detailed knowledge of the structure of members of this group, but reference is made to the importance of bacteria as **saprophytes** and **parasites**, and to their activities in the **nitrogen** and **carbon cycles**. There is also interest in the use of bacteria in **genetic engineering**.

KINGDOM PROTOCTISTA

The members of this kingdom are mostly small eukaryotic organisms, with membrane-bound organelles and a nucleus with a nuclear membrane. In this kingdom are found the organisms which are neither plants (developing from an embryo) nor animals (developing from a blastula) nor fungi (developing from spores). So the algae, including the seaweeds, the water moulds, the slime moulds and the protozoa, are all included in the Kingdom Protoctista. Most syllabuses include representatives of the following groups within the kingdom:

- **Phylum Rhizopoda**
  E.g. *Amoeba* – moves by means of pseudopodia; feeds heterotrophically by phagocytosis

- **Phylum Ciliophora**
  E.g. *Paramecium* – possess cilia; feeds heterotrophically; have two nuclei

In addition, some syllabuses require a knowledge of a flagellate, such as *Euglena*, and a parasitic protozoan, such as *Plasmodium*, the malarial parasite, or *Monocystis*, a parasite of earthworms.

> Representatives of these groups are found in most syllabuses

- **Phylum Oomycota**   Biflagellate spores produced; non-septate hyphae.
  E.g. *Phytophthora* – a parasite causing potato blight;
  Pythium – causing 'damping off' disease of seedlings

- **Phylum Chlorophyta**
  (Green algae): possess chlorophyll a and b as the main photosynthetic pigments; unicellular, filamentous or colonial.
  E.g. *Spirogyra* – simple unbranched filamentous alga, with spiral chloroplast, found floating in freshwater ponds

- **Phylum Phaeophyta**
  (Brown algae): photosynthetic pigments include chlorophyll and brown pigment, fucoxanthin.
  E.g. *Fucus* – common seaweed of the intertidal zone of rocky shores; possesses holdfast for attachment, stipe and lamina.

KINGDOM FUNGI

The members of this kingdom are eukaryotic, the body consisting of a network of threads called hyphae, forming a mycelium. There are no photosynthetic pigments present and feeding is heterotrophic; all members of the group being either saprophytic or parasitic. In some subgroups, the hyphae have no cross-walls, but in others cross-walls, or septa, are present. The spores lack flagella.

- **Phylum Zygomycota**
  No cross-walls except in connection with reproduction; asexual reproduction by means of spores produced in a sporangium; sexual reproduction by conjugation and zygospore formation.
  E.g. *Mucor* – pin mould: grows as a saprophyte on bread and dung.

- **Phylum Ascomycota**
  Cross-walls formed in this group; fusion of nuclei in sexual reproduction results in ascospore formation inside an ascus.
  E.g. *Saccharomyces* – yeast; a unicellular fungus which reproduces asexually by budding; can respire aerobically or anaerobically.

- **Phylum Basidiomycota**
  Cross-walls formed in this group; fusion of nuclei in sexual reproduction results in basidiospore formation in a basidium; basidia are found on a fruiting body composed of compacted hypae.
  E.g. *Agaricus* – common field mushroom: spore-producing layer protected by cap, or pileus, raised above the ground by a stipe, or stalk.

KINGDOM PLANTAE

The members of this kingdom are multicellular and photosynthetic. The cells are eukaryotic, have cellulose walls, vacuoles containing cell sap and chloroplasts containing photosynthetic pigments.

- **Phylum Bryophyta**
  The mosses and liverworts: alternation of haploid gamete-producing gametophyte with diploid spore-producing sporophyte; develops on the gametophyte.
  - **Class Hepaticae** – the liverworts: have simple thalloid body, unicellular rhizoids.
    E.g. *Pellia* – common, grows on banks of ditches and streams; spore capsule carried at end of long white seta.
  - **Class Musci** – the mosses: gametophyte divided into stem-like structure bearing leaf-like out-growths; rhizoids multicellular.
    E.g. *Funaria* – common moss of paths, greenhouses and bonfire sites.

CHAPTER 3 ESSENTIAL PRINCIPLES

In older classifications, the following phyla were grouped together to form the Phylum Tracheophyta; all the members possessing a well-defined sporophyte generation, vascular tissues made up of xylem and phloem and leaves with a waterproof cuticle.

> These phyla are often grouped together to form the phylum Tracheophyta

- **Phylum Filicinophyta**
  The ferns: show clear alternation of generations but with dominant sporophyte; gametophyte is a simple prothallus; sporangia in clusters.
  E.g. *Dryopteris* – common fern of woodlands with large divided fronds, underground rhizome and sporangia in clusters (sori)

- **Phylum Lycopodophyta**
  The club-mosses: small leaves arranged spirally around stem; sporangia in cones or strobili at apex of stem
  E.g. *Lycopodium* – on moorland and mountains, spores of one type only;
  E.g. *Selaginella* – two types of spore produced, microspores and megaspores, develop into male and female prothalli respectively

- **Phylum Sphenophyta**
  The horse-tails: vegetative stem bears whorls of small leaves at nodes; cones borne at apex of reproductive stem; spores of one type
  E.g. *Equisetum* – locally common on waste ground and in damp places

- **Phylum Coniferophyta**
  The conifers: reproductive structures borne in cones; ovules naked, no flowers or fruit
  E.g. *Pinus* – trees with needle-like leaves and cones

- **Phylum Angiospermophyta**
  The flowering plants: reproductive structures in flowers; ovules enclosed within ovary, which develops into a fruit after fertilisation
  - **Class Monocotyledons** – one seed leaf in the embryo, leaves with parallel veins, stem has scattered vascular bundles, flower parts in multiples of three
    E.g. Family Liliaceae: *Endymion* (Bluebell) – liliaceous monocotyledon
  - **Class Dicotyledons** – two seed leaves in the embryo, leaves with reticulate venation, vascular tissue in peripheral ring, flower parts in multiples of four and five.
    E.g. Family Ranunculaceae: *Ranunculus* (Buttercup) – polypetalous herbaceous dicotyledon;
    E.g. Family Labiatae: *Lamium* (Deadnettle) – sympetalous herbaceous dicotyledon

KINGDOM ANIMALIA

The members of this kingdom are multicellular and have heterotrophic nutrition.

- **Phylum Porifera**
  The sponges: body is a colony of unicells; skeleton made up of calcareous spicules; sessile; marine

- **Phylum Cnidaria**
  Sac-like body cavity, or enteron; radial symmetry; may exist in two forms, polyp and medusa, alternating in life cycle; stinging cells; two layers to body, ectoderm and endoderm
  - **Class Hydrozoa** – enteron sac-like, not divided up into compartments, in polyp form
    E.g. *Hydra* – freshwater hydroid, no medusoid form, solitary;
    E.g. *Obelia* – marine, colonial, has both polyp and medusoid forms; medusa free-swimming
  - **Class Anthozoa** – no medusa stage in this life cycle; enteron has been divided by radial partitions; sedentary; mouth opens into stomodeum
    E.g. *Actinia* – sea anemone: marine, solitary, ciliated planula larval stage in life cycle

- **Phylum Platyhelminthes**
  The flatworms: three layers to body, ectoderm, endoderm and mesoderm; flattened body; mouth but no anus; hermaphrodite
  - **Class Turbellaria** – free-living flatworms, the planarians
    E.g. *Planaria* – common in freshwater; ciliated on underside; lives under stones, carnivorous
  - **Class Trematoda** – the flukes: parasitic; suckers for attachment to host
    E.g. *Fasciola hepatica* – lives in liver of sheep; has a number of different larval forms; secondary host in its life cycle is a snail
  - **Class Cestoda** – the tapeworms: no gut; suckers and hooks for attachment; body consists of proglottides
    E.g. *Taenia solium* – the pork tapeworm: primary host man, secondary host pig; lives in small intestine of man and absorbs digested food

- **Phylum Nematoda**
  The roundworms: unsegmented body; cylindrical; pointed at both ends; mouth and anus present; body covered with thick elastic cuticle
  E.g. *Ascaris* – parasitic roundworm found in small intestine of man and pig

- **Phylum Annelida**
  The segmented worms: body made up of segments; chaetae; show metameric segmentation, i.e. repetition of structures along length of body
  - **Class Polychaeta** – the bristle worms: marine; separate sexes; lots of chaetae on parapodia
    E.g. *Nereis* – the rag worm: carnivorous, free-swimming, with well-developed head;
    E.g. *Arenicola* – the lug worm: burrowing type, not such an obvious head
  - **Class Oligochaeta** – earthworms: few chaetae, no parapodia, hermaphrodite, head not obvious.
    E.g. *Lumbricus* – lives in soil, four pairs of chaetae per segment

- **Phylum Mollusca**
  Large phylum of soft-bodied animals; head well developed with tentacles and eyes; muscular foot; dorsal visceral mass; mantle covers visceral mass and secretes shell; gills
  - **Class Gastropoda** – snails and slugs; body undergone torsion during development; visceral mass coiled into a spiral; single shell
    E.g. *Helix aspera* – the garden snail: herbivorous, rasping plant material with a radula, gills absent, uses cavity like a lung
  - **Class Pelycopoda** – bivalve molluscs: head and radula absent; gills present; ciliary feeders; shell in two halves; wedge-shaped foot
    E.g. *Mytilus* – the marine mussel: sessile attached to rocks, feeds on detritus and plankton

- **Phylum Arthropoda**
  Animals with jointed appendages; chitinous cuticle forming exoskeleton
  - *Class Crustacea* – crabs, prawns, woodlice: head and thorax form cephalothorax; head bears two pairs of antennae; compound eyes
    E.g. *Daphnia* – water flea: four pairs of feathery limbs used for filter feeding and gas exchange; antennae used for locomotion;
    E.g. *Carcinus* – shore crab: four pairs of jointed walking legs; one pair of pincers, thick carapace, abdomen reduced, stalked eyes
    E.g. *Oniscus* – woodlouse: terrestrial; found under logs; feeds on decaying organic matter; abdominal appendages modified for gas exchange
  - **Class Chilopoda** – centipedes: terrestrial; body divided into head and trunk; simple eyes; one pair of antennae; one pair of legs per segment.
    E.g. *Lithobius* – garden centipede: one pair of legs per segment; carnivorous, with poison claws; simple eyes
  - **Class Insecta** – the insects: body divided into head, thorax and abdomen; one pair of antennae; two pairs of wings; three pairs of jointed legs; compound eyes; tracheal system for gas exchange
    Hemimetabola – egg, nymph, adult, e.g. dragonflies, locusts, earwigs
    Holometabola – egg, larva, pupa, adult, e.g. butterflies, moths, bees, ants

CHAPTER 3 ESSENTIAL PRINCIPLES

- **Class Arachnida** – the spiders: body divided into prosoma and opisthosoma; no antennae; four pairs of walking legs on prosoma.
  E.g. *Epeira* – garden spider: feeds on small animals trapped in its web

■ **Phylum Echinodermata**
Spiny-skinned animals; marine; pentaradiate; possess tube feet; skin with calcareous ossicles and spines; mouth on lower side; anus on upper side
E.g. *Asterias* – common starfish: has five arms with tube feet on lower surface opening to water vascular system, the madreporite, on the upper surface

■ **Phylum Chordata**
Possess a notochord at some stage in the life cycle – a dorsally situated rod of cartilage-like material; also possess a dorsal, tubular nerve cord; post-anal tail; segmented muscle blocks; pharynx perforated by paired clefts (pharyngeal clefts); closed blood vascular system
- **Sub-phylum Vertebrata/Craniata** – the vertebrates: in this group the notochord is replaced by the vertebral column in the adult, and the brain is encased in a cranium
- **Class Chondrichthyes** – cartilaginous fish: skeleton made of cartilage; fleshy fins; ventrally situated mouth; paired gill slits.
  E.g. *Scyliorhinus* – the dogfish: found in coastal waters; feeds on crustaceans and small fish; eggs are known as mermaid's purses
- **Class Osteichthyes** – the bony fish: skeleton made of bone; gills covered by operculum; fins supported by rays; mouth terminal.
  E.g. *Clupea* – the herring: occur in shoals; carnivorous; of economic importance as food
- **Class Amphibia** – possess simple lungs; no scales; moist skin; larval form with gills; adults usually terrestrial; larval forms aquatic and undergo metamorphosis into adults; teeth, if present, all one type (homodont),
  E.g. *Rana* – the frog: occurs in damp vegetation; feeds on insects and small molluscs; hibernates in winter; mates in spring, producing familiar frog spawn
- **Class Reptilia** – possess dry scaly skin; lungs; lay eggs enclosed in a shell (cleidoic); no larval forms; teeth – homodont.
  E.g. *Lacerta* – the lizard: found in sand dunes; feeds on spiders and insects; eyes with three eyelids; wide mouth and long tail
- **Class Aves** – birds: possess feathers; forelimbs are wings; horny beak; no teeth present; lungs, cleidoic eggs; endothermic.
  E.g. *Columba* – the pigeon: found in urban areas; strong fliers; large eyes for keen sight; feet adapted for walking and perching; streamlined shape for flying
- **Class Mammalia** – possess mammary glands which secrete milk to feed young; skin covered in hair or fur; endothermic; lungs; teeth – heterodont (more than one type); limbs held under body; viviparous – no eggs laid usually

Very few syllabuses specify particular mammals for study, though many do make reference to humans in other sections. The relevant syllabus should be checked carefully and if further information on mammals, or any other groups of organism, is required, reference can be made to one of the texts mentioned in the bibliography on this page.

PRACTICAL WORK

> Careful drawings, with annotations, can help

There is a great deal of practical work attached to this section of any syllabus. It is important to be able to recognise characteristic features of organisms, so that they can be placed in their correct taxonomic groups, and to be able to pick out features which adapt organisms to their habitat. The best method of familiarising oneself with these features is to make careful drawings of specimens, annotating points of biological significance. The drawings do not need to be artistic, but should have clear outlines and accurate proportions. Each drawing should have a scale ($\times\frac{1}{2}$, $\times 1$, $\times 3$, depending on its size) indicated on the drawing.

The use of keys has already been mentioned, and it is relevant here to stress the importance of being able to pick out specific features of an organism in order to be able to identify it successfully. This requires practice and perseverance, and it is probably better to begin with a known organism before trying an unknown one.

When making keys, a sound knowledge of the important features of the organisms involved is invaluable, as it helps to pick out those which are significant. Obvious structural features are the best to choose, and most keys should be dichotomous: i.e. divide the organisms into two groups at each stage. An example of a dichotomous key is given at the end of the chapter.

Further reading
Jenking M and Boyce A, 1979, *Diversity of Life*. Macmillan Education
Miller R N, 1982, *Plant Types 1*. Hutchinson
Miller R N, 1985, *Plant Types 2*. Hutchinson
Robinson M A and Wiggins J T, 1971, *Animal Types 1. Invertebrates*. Hutchinson
Robinson M A and Wiggins J T, 1971, *Animal Types 2. Vertebrates*. Hutchinson

EXAMINATION QUESTIONS AND ANSWERS

As has already been mentioned, most of this section of any syllabus will be examined either as part of a practical examination or as part of the assessed work carried out during the course. Questions may be set on theory papers, but it would be most unusual for a long essay question to be set. Most questions chosen to illustrate this section have therefore been taken from short-answer papers or from practical examinations. You can see student answers to these questions together with examiner comments.

STUDENT ANSWERS WITH EXAMINER COMMENTS

Question 3.1

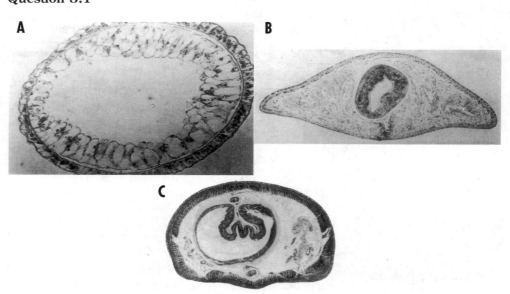

Photographs A, B and C are of transverse sections of animals belonging to three different phyla: Coelenterata (Cnidaria), Platyhelminthes and Annelida. Give **two** structural features *shown in each photograph* which help to identify the phylum of each animal.

> **Both acceptable features**

A Phylum: Coelenterata (Cnidaria)
Structural feature 1 — Diploblastic
                  2 — Radially symmetrical

B Phylum: Platyhelminthes

> **Correct point**

> **Acceptable**

Structural feature 1 — Triploblastic
                  2 — Flattened shape

> **Both acceptable structural features**

C Phylum: Annelida
Structural feature 1 — Triploblastic
                  2 — Coelomate

(Total 6 marks)
(AEB)

Question 3.2

The drawings show the anterior segments of three different insect larvae. All three larvae belong to the same insect order, the Lepidoptera (butterflies and moths).

species A — thorax, simple eyes

species B

species C — antenna

0 mm 10

a) i) Construct a table comparing **six** features of the anterior segments of the three larvae. (6)
   ii) Construct a dichotomous key enabling the three larvae to be distinguished from one another. (3)
b) In what ways are the features shown in the drawings above and to the left adaptations to the different modes of life of larval and adult stages of the life cycle? (9)
c) What are the likely advantages of a life cycle with different larval and adult stages? (3)

(Total 21 marks)
(AEB)

3.2 a) i)

	A	B	C
Setae	small on head, pair on each segment	none	many on head, quite long. Several on each segment
Thorax	3 definite segments	segments not clearly defined	3 definite segments
Eyes	5 simple eyes present on each side of head	5 simple eyes present on each side of head	5 simple eyes present on each side of head
Antennae	pair – reduced	pair – very reduced	pair – large
Spiracles	pair on 1st segment of thorax	pair on 1st segment of thorax	pair on 1st segment of thorax
Legs	3 pairs of walking legs on thorax	3 pairs of walking legs on thorax	3 pairs of walking legs on thorax

> *Good feature to pick out, but description could be less complex; setae could be called 'hairs'*
> *Acceptable*
> *Acceptable – a similarity not a difference*
> *Acceptable*
> *Another similarity*
> *A reasonable attempt but more emphasis on differences needed. As a scale is provided, measurements of head, legs could be performed*

ii) 1. a) Large dark 'eye spot' marking on first abdominal segment (specimen B)
   b) No 'eye spot' marking (2)
2. a) Many long setae all over head (specimen C)
   b) Shorter setae around mouth (specimen A)

> *A workable key, dichotomous. Other features could have been used to construct the key*

b) The larvae all have small simple eyes, ocelli, which only perceive light and dark. As they only have a slow mode of life eating vegetation, they do not need accurate vision to survive.

The adult has large, compound eyes able to perceive movements accurately and in colour. The adult has to see well to fly and to find food, coloured flowers, and a mate.

The adult has a long coiled proboscis for reaching nectaries deep in flowers for food. The larvae have completely different cutting mouthparts for dealing with leaves.

The larvae of species B has warning colouration in the form of a large 'eye spot' marking. This is to make the larva appear bigger and more dangerous to possible predators, so that they are less likely to attack. Species C has setae all over, as a defence mechanism, so it may be dropped if picked up by a bird, etc.

The three pairs of true legs, shown by the larvae, all have hooked claws. These assist with gripping on to the smooth surfaces of leaves.

The adult has long, jointed clubbed antennae. These hold mechanoreceptors and chemoreceptors. The length means scents from the air and vibrations can be detected sooner in flight and from all directions. The antennae of the larvae are much shorter and less directional, only sensing around the mouth area.

> *A number of appropriate points have been clearly made*

> **Good points made here**

> **Perhaps reference could be made to survival of the species as well**

```
c) If a species has several stages in the life cycle, the
larval stage may be able to disperse the species further
before maturing. The larval stage may be the eating and
growing stage, whilst the adult can reproduce: e.g.
butterfly.
   The larval and adult stages are likely to eat
differently and colonise different ecological niches,
therefore there will be less intraspecific competition for
food, O₂, etc.
```

Question 3.3

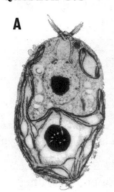
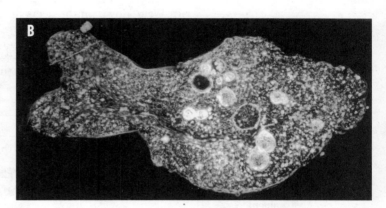

Photograph **A** shows a section through a motile green unicellular organism.
Photograph **B** shows a whole motile unicellular organism.

a) Using evidence visible in the photographs, list **five** important differences between the structure of organisms **A** and **B**.

> **Correct for A, but B not comparable**

> **Correct answers**

A	1) Possesses 2 flagella	B	1) No flagella
A	2) Has a fixed cell wall	B	2) Has a movable cell membrane

> **Better to refer to lack of cell wall**

> **Correct for A and B**

> **Correct for A and B**

> **Correct for A and B**

A	3) Possesses a chloroplast	B	3) No chloroplast
A	4) No contractile vacuole	B	4) Possess a contractile vacuole
A	5) No food vacuoles	B	5) Possesses food vacuoles

b) In terms of their modes of nutrition, give one word to describe organism **A** and one word to describe organism **B**.

> **Both correct**

A   Autotrophic   B   Heterotrophic   (2)

c) i) Name the class to which organism B belongs.

> **Correct answer**

Sarcodina   (1)

ii) Sometimes in the past organisms such as **A** have been classified as Protozoans. Which protozoan class would best suit organism **A**?

> **Correct**

Flagellata   (1)

d) Name **two** different types of carbohydrate molecule that could be present in organism **A** in substantial quantities.

> **Correct**

Starch   Glucose   (2)

e) Which organism would you consider to be more advanced? Give a reason for your answer.

CHAPTER 3 THE DIVERSITY OF ORGANISMS

Organism B is more advanced. An amoeba has an osmoregulatory organ in the form of a contractile vacuole. It also shows some sensitivity e.g. retracting pseudopodia to touch, chemicals, heat, for protection. (2)

(Total 13 marks)
(ODLE)

> In (a) other features to which reference could be made are eyespot in A, not in B; pseudopodia in B, but not in A

> This could be a matter of opinion! The first reason given is not a good one. The most acceptable answer would be organism A. Organism A can synthesise its own food materials and locomote more rapidly. It is also sensitive to light.

Question 3.4

a) i) Name a class of the phylum Annelida.

> Correct answer

Oligochaeta (1)

ii) Write down, in the accepted scientific manner, the name of a species of annelid from this class.

> Correct answer: genus with capital letter, species name with lower case

Lumbricus terrestris (2)

b) i) State two characteristics that enable you to classify the organism into the phylum Annelida.

> Correct features

Bilaterally symmetrical
Metameric segmentation (2)

ii) State one characteristic of the class of Annelida you have chosen.

> Correct. Acceptable alternatives would be no parapodia, few chaetae per segment

Indistinct head without appendages (1)

c) Describe the mechanism of locomotion in a named annelid.

Name of annelid   Earthworm

There is no hard skeleton, instead there is a fluid under pressure = hydrostatic skeleton. Circular and longitudinal muscles in the body wall contract against the coelomic fluid in the body cavity. The transverse septa, dividing the body into segments, prevent the fluid moving up and down the body; therefore localised bulges can occur from muscle action, locomotion is achieved by these bulges being formed in waves down the body. (4)

(Total 10 marks)
(ODLE)

> Reasonable answer, but reference to the action of the chaetae gripping the soil is needed, together with more precise details of which segments get short and fat, and which long and thin

Question 3.5

The diagram classifies living organisms into five kingdoms.

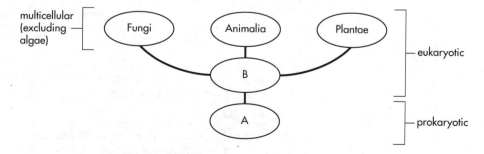

a) Name kingdoms **A** and **B**.

> Alternative acceptable name: Prokaryotae

A   Monera

> Now more accurately Protoctista but Protista acceptable

B   Protista

b) Consider the three kingdoms Fungi, Animalia and Plantae.

Give **one** feature of each which distinguishes it from the other two.

Fungi	Cell walls present, not of cellulose
Animalia	Nervous co-ordination ← "Correct answers"
Plantae	Feed by photosynthesis

(Total 5 marks)
(AEB)

> 66 Can use this, but lack of photosynthetic pigment or reference to hyphae and mycelium more appropriate 99

> 66 Better to use 'possession of photosynthetic pigments' or 'chlorophyll' rather than the process 99

Question 3.6
The diagrams shows one way of classifying living organisms.

Prokaryotae — (A) __Protozoa__ — Fungi — Animalia — Plantae

Under Protozoa: Euglenophyta, Rhizopoda, Chlorophyta

Under Animalia: Cnidaria, Platyhelminthes, Annelida, Arthropoda, (B) __Vertebrata__

Under Vertebrata: Osteichthyes, Avia (Aves), Mammalia

> 66 Incorrect. The kingdom is Protoctista. Protozoa is no longer used and would be incorrect anyway as Chlorophyta would not be included in Protozoa 99

> 66 Incorrect. Vertebrata is a sub-phylum. This should be Chordata 99

a) Complete the diagram by filling in spaces **A** and **B**. (2)
b) Describe **one** way by which you could distinguish a member of the phylum Euglenophyta from a member of the phylum Rhizopoda.

   It has a flagellum _____ (1)

c) From the diagram:
   i) name **one** group containing organisms which are diploblastic and radially symmetrical

   Cnidaria _____

   ii) name **one** group containing organisms which are eukaryotic but do *not* possess undulipodia

   Arthropoda _____ (2)

(Total 5 marks)
(AEB)

> 66 Correct, but perhaps too brief. 'It has one or several flagella' would be better. The term undulipodium could be used instead of flagellum 99

> 66 Correct 99

> 66 Correct. Undulipodia are eukaryotic cilia and flagella which are not possessed by the Arthropoda 99

NB: Some Review Questions on this chapter can be found at the end of Chapter 4.

GETTING STARTED

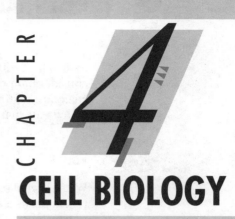

CHAPTER 4
CELL BIOLOGY
MOLECULES
CELLS AND ORGANELLES
ENZYMES
PRACTICAL WORK

A knowledge of cell biology is fundamental to an understanding of the physiological processes of living organisms.

The structure and properties of biologically significant molecules, such as **water**, **carbohydrates**, **lipids**, **proteins** and **nucleic acids**, are important in gaining an appreciation of such metabolic processes as **photosynthesis**, **respiration** and **protein synthesis**.

Cells are considered to be the basic units which make up living organisms – a concept put forward by Schleiden and Schwann in 1839 in their Cell Theory. The contents of cells can be studied using the light microscope; even more information can be gained from electron micrographs of plant and animal cells which reveal the presence and structure of sub-cellular **organelles**.

In multicellular organisms, **cells** are aggregated into **tissues**, **tissues** form **organs**, **organs** form part of **organ systems** and an **organism** consists of a number of co-ordinated **organ systems**.

All the **metabolic reactions** of living organisms are controlled by biological catalysts called **enzymes**, which are globular protein molecules. A knowledge of the properties of enzymes and the conditions affecting the way in which they function is also necessary in order to understand how these metabolic reactions proceed.

CHAPTER 4 CELL BIOLOGY

ESSENTIAL PRINCIPLES

There are many different approaches to the biology of the cell, but it is proposed here to deal first with biologically important molecules and protein synthesis, followed by a consideration of cells and organelles, building up into tissues and organs. Finally, the structure and mode of action of enzymes will be considered together with the way in which external factors affect their activities.

MOLECULES

WATER

Water is vital to all living organisms, as a constituent of cells, as a reactant in metabolic reactions, as a solvent and, for a large number of aquatic organisms, as a habitat. It has unusual properties for a molecule of its size and many of these have biological significance.

> Some important properties of the water molecule

Polar molecule

Water is a polar molecule, with a slight positive charge at one end of the molecule and a slight negative charge at the other end. The oxygen atom has a negative charge and attracts the positively charged hydrogen atoms of other water molecules. Hydrogen bonds are formed between the water molecules, holding them together; see Figs 4.1a) and 4.1b).

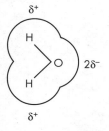

δ^+ and δ^- represent small charges

Fig. 4.1a) Polarity and bonding in water molecules

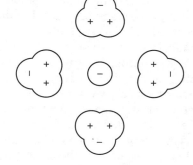

Fig 4.1b) The orientation of the water molecules around positively and negatively charged ions

Solvent

Because of its polarity, water is a very good solvent for other polar molecules and ionic substances which become more reactive in solution, so it is the medium for metabolic reactions in cells. Water is vital in the transport of substances around the bodies of organisms; this is seen clearly both in the blood (plasma is made up of over 90 per cent water with soluble food materials, hormones and urea dissolved in it) and the vascular tissues in plants (phloem transport of sucrose and amino acids in solution).

Heat capacity

Water has a high heat capacity, which means that it takes a large amount of heat energy to cause a small rise in its temperature, so that temperature changes in water are usually quite small. This is important inside cells, where metabolic reactions are enzyme-controlled, and also externally, in the provision of a fairly constant environment for aquatic organisms.

LATENT HEAT OF VAPORISATION

Water also has a high latent heat of vaporisation, which means that a great deal of heat energy is required to change it from its liquid state to a vapour; this is reflected in the unusually high boiling point for the size of the molecule. The biological significance of this is seen in sweating and panting in mammals, valuable methods of temperature control where the heat used for the vaporisation of the water in sweat is lost from the body, thereby cooling it down.

DENSITY

Water has its maximum density at 4°C, so when it is cooled below this temperature its density decreases and ice floats on the surface. This has been important in the evolution of life on this planet, particularly in the Arctic and in temperate regions with cold seasons.

SURFACE TENSION

Water has a high surface tension, the individual molecules having a great attraction for one another. This high cohesion of water molecules is important in the movement of water through xylem tissue in plants. The high surface tension is used by some small aquatic organisms for support.

HYDROLYSIS REACTIONS

Water is an important reactant in hydrolysis reactions, such as those involved in digestion, and in photosynthesis, where it is a source of hydrogen.

CARBOHYDRATES

Carbohydrates are organic molecules containing carbon, hydrogen and oxygen; they are divided into:

- monosaccharides
- disaccharides
- polysaccharides

Monosaccharides

The monosaccharides, or **simple sugars**, have a general formula $(CH_2O)_n$, where n can be a number from 3 to 9. The most common monosaccharides are the trioses, where n is 3; the pentoses, where n is 5; and the hexoses, where n is 6. The triose sugars are important in metabolism (see respiration and photosynthesis), the pentoses include ribose and deoxyribose, involved in the synthesis of nucleic acids, and the hexoses are the main sources of energy for living cells.

- **Glucose** is the most common hexose sugar and is the most common respiratory substrate. It exists in two forms, α glucose and β glucose, which differ only in the way in which the H and OH groups are attached to one of the carbon atoms (see Fig. 4.2).

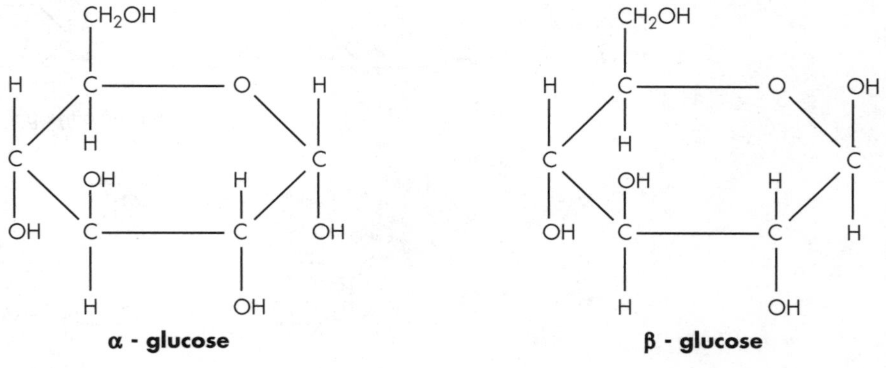

66 Triose sugars are important in metabolism 99

Fig. 4.2 Structure of glucose

α - glucose   β - glucose

Disaccharides

Two monosaccharide units linked together by the formation of a **glycosidic** bond with the elimination of a molecule of water results in the formation of a disaccharide molecule (see Fig 4.3). This type of reaction is known as a condensation reaction, and is common in the formation of polysaccharides from monosaccharide monomers, and in the formation of polypeptide chains from amino acids, where peptide links are formed.

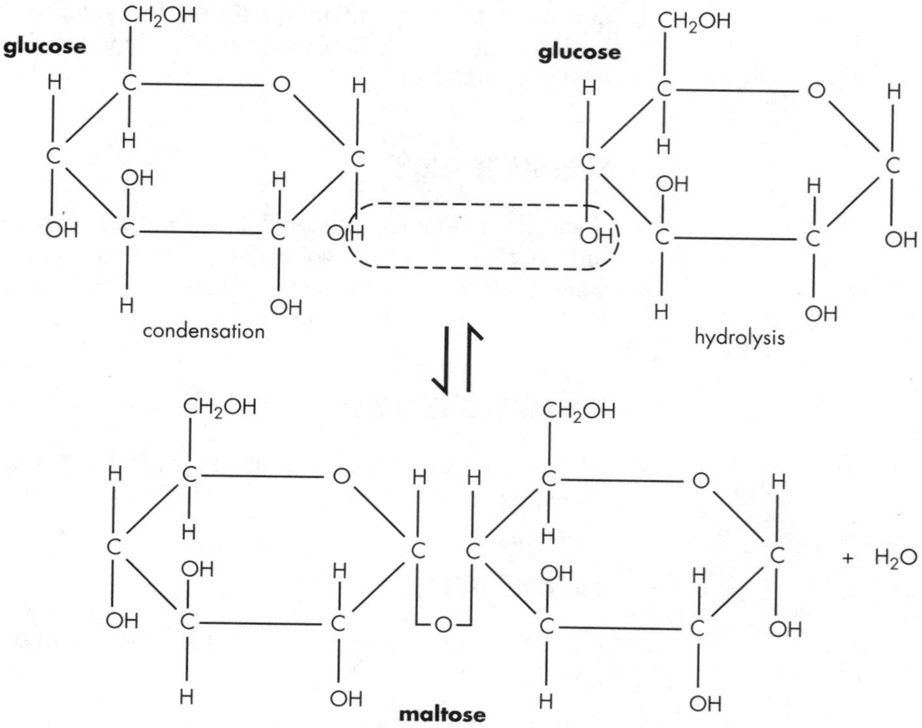

Fig. 4.3 Condensation of two glucose molecules to form the disaccharide maltose

Polysaccharides

Polysaccharides are formed from very large numbers of monosaccharide units linked together, and they are important chiefly as stores of energy and as structural molecules.

- **Starch** and **glycogen**, built up of α-glucose molecules, are energy stores in plants and animals respectively. They are well suited to their function as they are insoluble, compact molecules and are thus unable to alter the osmotic properties of the cells in which they are stored (see Fig. 4.4a)).
- **Cellulose** is a structural molecule, consisting of long chains of β-glucose molecules linked together, and linked to other long chains by **hydrogen bonds**. These chains are grouped together into **microfibrils**, which have enormous tensile strength and which are laid down in layers forming the cellulose cell walls in plants (see Fig. 4.4b)). Cellulose is of great economic importance as a constituent of paper and in the manufacture of fabrics.

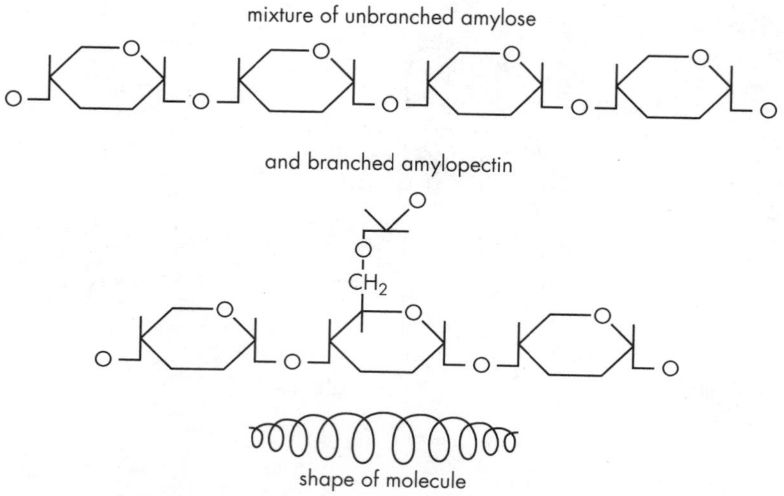

mixture of unbranched amylose

and branched amylopectin

shape of molecule

Fig. 4.4a) Starch molecule structure

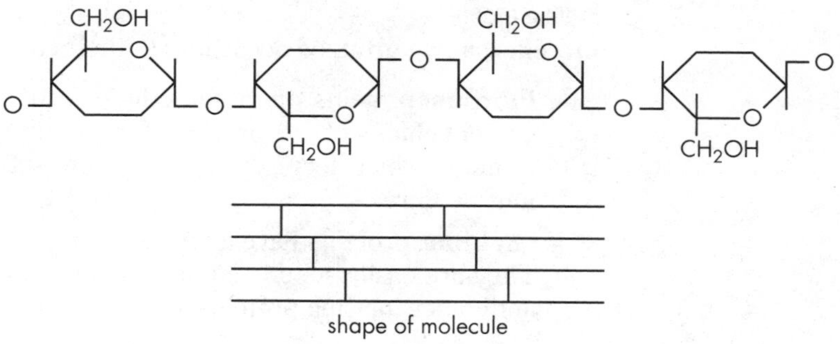

Fig. 4.4b) Cellulose molecule structure

LIPIDS

Lipids are **triglycerides**, composed of carbon, hydrogen and oxygen, but the oxygen content is very low. They are formed by condensation reactions between **glycerol** and **fatty acids**, and their properties depend on the nature of the fatty acids involved in their formation. They are **non-polar** molecules and so are insoluble in water. Their major function is as energy stores, having higher energy yields when completely oxidised than an equivalent mass of carbohydrate. They are good heat insulators; they provide buoyancy and form protective cushions in animals.

- **Phospholipids** are lipids which contain a phosphate group. The phosphate group makes that part of the molecule polar and therefore soluble in water, while the fatty-acid chains are non-polar. Phospholipid molecules are important in the formation and functioning of the membranes in cells (see Fig. 4.5).

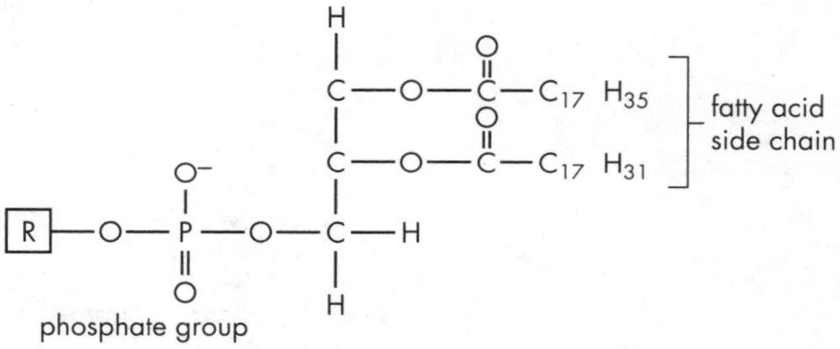

Fig. 4.5 Structure of a phospholipid

PROTEINS

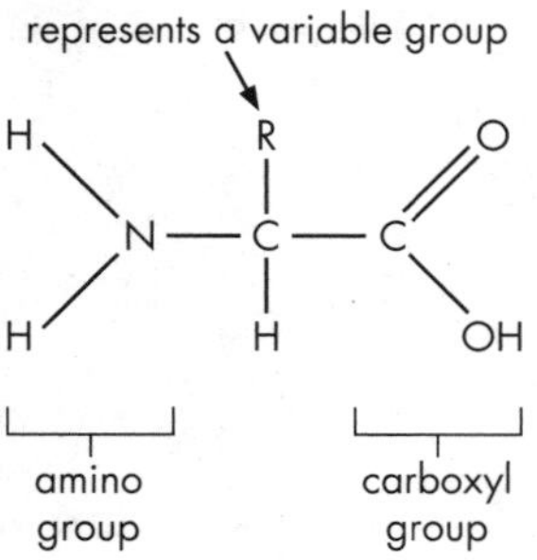

Fig. 4.6 Structure of an amino acid

Proteins are polymers of **amino acids**, and are important in the formation of living organisms. Proteins are made up of carbon, hydrogen, oxygen and nitrogen atoms, together with phosphorus and sulphur. Amino acids possess an **amino** group, NH_2-, at one end of the molecule, and a **carboxyl** group, $-COOH$, at the other.

Properties
They are crystalline and colourless, and are usually soluble in water, where they form **zwitterions** with a positive charge at the amino end of the molecule and a negative charge at the carboxyl end. They are **amphoteric** and so can act as buffers, resisting changes in pH, by donating hydrogen ions if the pH increases and accepting them if the pH decreases.

Structure

> The structure of a protein

The **primary structure** of a protein is the sequence of amino acids joined together to form a **polypeptide** chain. This chain is usually coiled in an α-helix, forming the **secondary structure**, and then folded into a specific shape known as the **tertiary structure**. Both the α-helix and the tertiary structures are maintained by bonds: **ionic**, **hydrogen** or **disulphide**. Many proteins consist of more than one polypeptide chain, and the joining and orientation of these chains is referred to as the **quaternary structure**.

Protein groups
Proteins can be grouped according to their structure: fibrous or globular.

- **Fibrous** proteins often show little tertiary structure, being composed of long parallel chains of polypeptides forming fibres or sheets. These are the structural proteins which form the basis of connective tissues, tendons, bone matrix and muscle fibres.
- **Globular** proteins have a tertiary structure and are folded into spherical shapes. They are easily soluble and can form colloidal suspensions. They form enzymes, antibodies, plasma proteins and hormones (see Fig. 4.7).
- **Conjugated** proteins have non-protein groups associated with them, known as prosthetic groups. Haemoglobin is a conjugated protein, consisting of a pigment, haem, associated with protein.

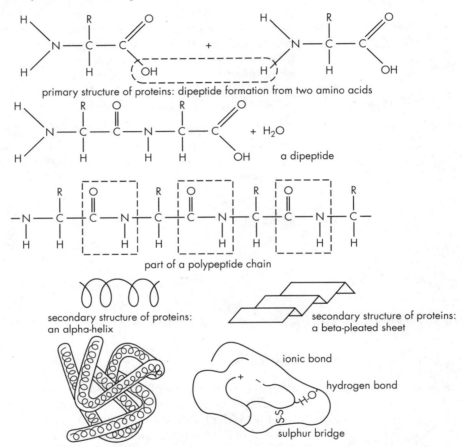

Fig. 4.7 Structure of proteins

NUCLEIC ACIDS

Nucleic acids make up the genetic material of all living organisms. Their structure was worked out by Watson and Crick in the 1950s.

Structure

> **The structure of nucleic acids**

They are built up of units called **nucleotides**, which are formed by condensation from a 5-carbon sugar, an organic base and a phosphate group.
The sugars are as follows:

- in **ribonucleic acid (RNA)** the sugar is ribose
- in **deoxyribonucleic acid (DNA)** the sugar is deoxyribose
- both sugars are pentoses with five carbon atoms

The bases are two types, **purines** and **pyrimidines**

- the purines are **adenine** and **guanine**
- the pyrimidines are **thymine, cytosine** and **uracil.**

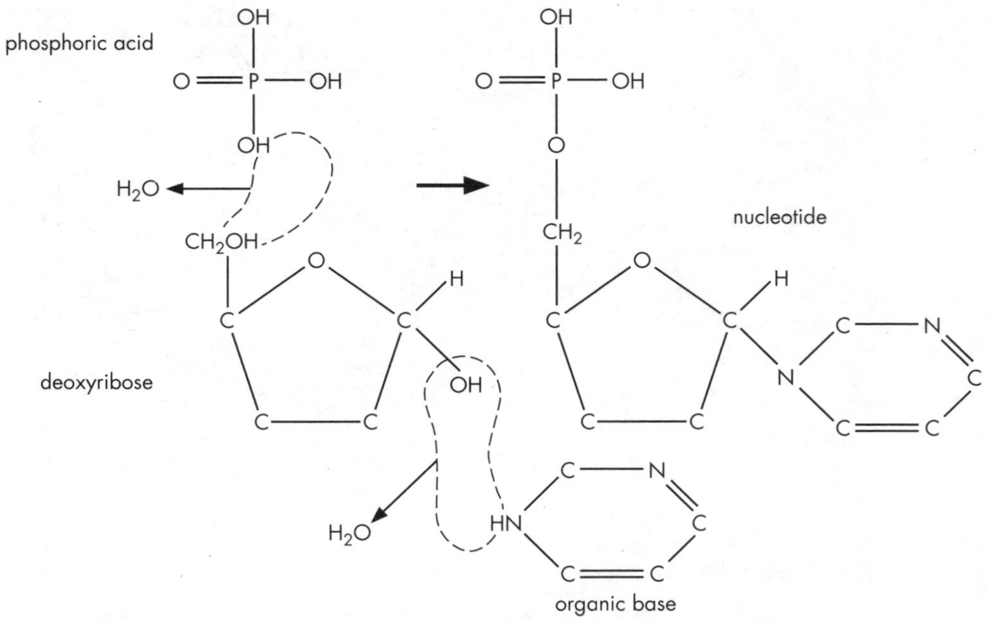

Fig. 4.8 Formation of a nucleotide

- DNA nucleotides contain adenine, guanine, thymine or cytosine
- RNA nucleotides contain uracil instead of thymine.

Nucleotides are linked together by condensation to form nucleic acids; RNA consisting of a single strand and DNA consisting to two strands held together by hydrogen bonds between the complementary bases, coiled helically (see Fig. 4.9).

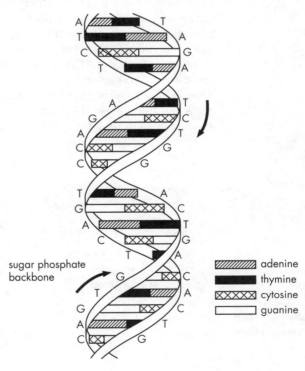

Fig. 4.9 Structure of a DNA molecule

Role of DNA

Avery and Griffiths showed that the role of DNA in cells is to carry the genetic information; it is the molecule of inheritance and as such it must be capable of replication, so that exact copies of the information are passed on to daughter nuclei during nuclear division (see Fig. 4.10).

The work of Beadle and Tatum, using the fungus *Neurospora*, indicated that genes controlled the formation of enzymes, and put forward the **'one gene – one enzyme hypothesis'**, which was later modified to **'one gene – one polypeptide'**. The hypothesis suggests that a portion of the DNA molecule carries the coded information for the synthesis (by ribosomes in the cytoplasm) of one polypeptide chain. The code

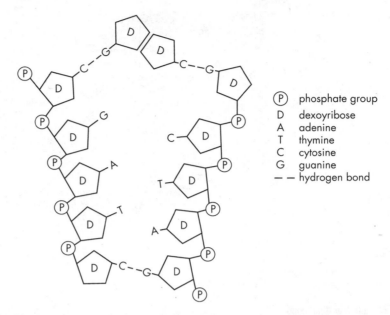

Fig. 4.10 Replication of DNA

is in the form of triplets of bases in the DNA, each triplet coding for one of the amino acids in the polypeptide.

The most recent terminology refers to the portion of a DNA molecule carrying the coded information for the synthesis of one polypeptide chain as a **cistron**. The information is in the DNA in the nucleus of the cell and the site of synthesis is the ribosome in the cytoplasm. The information on the DNA is copied by the formation of a **messenger RNA** molecule in a process called **transcription**. A portion of the DNA unzips and the molecule of RNA is built up alongside the unzipped portion, the base sequence being a complementary copy of that on the DNA. The messenger RNA molecule moves from the nucleus to the cytoplasm, where it attaches itself to a **ribosome**. **Transfer RNA** molecules bring specific amino acids to the ribosome and these are assembled in the correct sequence, according to the triplet sequence on the RNA; this is the process of **translation**. Peptide bonds are formed between the amino acids and the polypeptide chain is built up. Several ribosomes may become attached to a molecule of mRNA; the whole structure is then known as a **polysome** (see Fig. 4.11).

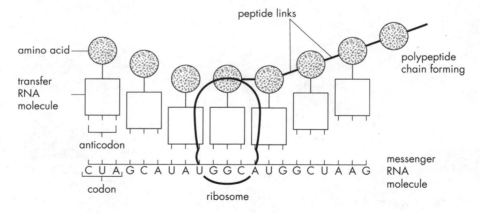

Fig. 4.11 Polypeptide formation in protein synthesis

CELLS AND ORGANELLES

In the introduction to the chapter we referred to **cells** as the basic units of living organisms. The cells of plants and animals have common features, such as:

- cell membranes
- cytoplasm
- nucleus
- mitochondria

They also have important differences, which should be carefully noted.

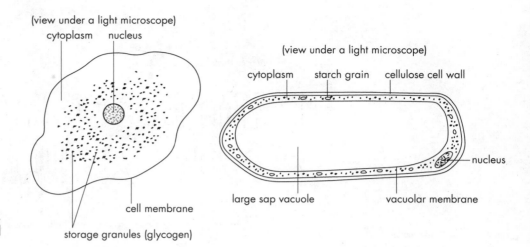

Fig. 4.12a) A typical animal cell, and b) A typical plant cell

Microscopy and cell structures

The structures shown in the diagrams can be observed using a light microscope, but **electron microscopy** enables the fine structure of cells and their organelles to be seen (see Fig. 4.13). The illustrations show typical cells, but the majority of cells forming the tissues and organs of organisms become highly specialised for particular functions: e.g. neurones, muscle cells. These will be dealt with in more detail in the relevant sections.

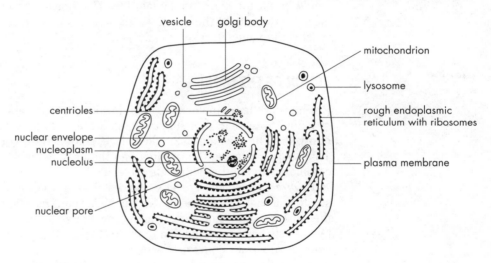

Fig. 4.13 Electron microscope diagram of a generalised animal cell

In addition to a knowledge and understanding of the structure and functioning of the organelles within a cell, it is also important to have an understanding of the **membranes** involved in cells:

> Study the membranes involved in cells

- the **cell** membrane or **plasma** membrane surrounding the cell contents
- the membranes of the **endoplasmic reticulum**
- the membranes which surround organelles such as **mitochondria** and **chloroplasts**

FLUID MOSAIC MODEL

The most recent hypothesis for the structure of the cell membrane is the Fluid Mosaic Model, put forward by Singer and Nicholson in the 1970s. This structure is thought to apply to biological membranes in general and is supported by evidence from electron microscope studies (see Fig. 4.14).

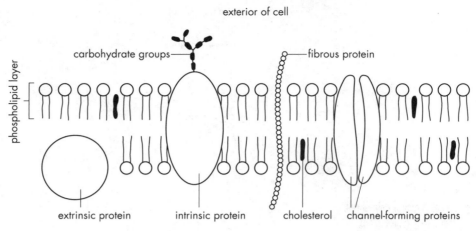

Fig. 4.14 Fluid-mosaic model of cell membrane

ENZYMES

Enzymes are **protein** molecules, made by living cells, which act as **catalysts** and speed up the rate of metabolic reactions, by lowering the **activation energy** required. In the absence of enzymes, the **anabolic** and **catabolic** reactions that make up **metabolism** would be far too slow to maintain life. The activation energy is the energy needed to increase the rate of the reaction; glucose has to be made more reactive before it will react with oxygen in respiration.

ACTION OF ENZYMES

Enzymes are large molecules. Each has its own special shape, with an area, the **active site**, on to which the substrate molecules bind, forming an **enzyme-substrate complex**. The reaction then takes place, an **enzyme-product complex** being formed. This splits, releasing the product and the enzyme, which is available to form another complex with another substrate molecule.

Enzyme molecules are **specific**, catalysing one, or one type of, reaction and it is suggested that this is due to the particular configuration of the active site into which the substrate molecules fit like a key, giving rise to the **lock and key hypothesis**. This hypothesis has been modified to the **induced fit hypothesis**, where it is thought that when a substrate combines with an enzyme, it induces the enzyme structure to fit, moulding the amino acids of the active site into the right configuration in order to carry out the reaction (see Fig. 4.15).

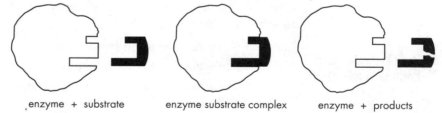

Fig. 4.15 Action of an enzyme

Cofactors

Most enzymes need another non-protein **cofactor** to be present. A cofactor can be an **inorganic ion**, a **prosthetic group** or a **coenzyme**. Inorganic ions are known as enzyme **activators**, and a good example is shown by the increased activity of salivary amylase in the presence of chloride ions. A prosthetic group is an organic molecule, such as **haem** in the **cytochromes**, which acts as an **electron carrier**. **NAD** and **NADP** are coenzymes derived from vitamins and they act as links between two different enzyme systems. Reference will be made to these later in the chapters dealing with photosynthesis and respiration.

CLASSIFICATION OF ENZYMES

> A useful classification of enzymes

Enzymes are classified according to the reactions that they catalyse, and the following is a list of the groups which are commonly encountered in A-level Biology.

- **Transferases** — Catalyse the transfer of atoms or groups from one molecule to another.
- **Dehydrogenases** — Catalyse the removal of hydrogen atoms.
- **Oxidases** — Catalyse the addition of oxygen to hydrogen.
- **Hydrolases** — Catalyse the addition of water or break bonds by the addition of water.
- **Decarboxylases** — Catalyse the removal of CO_2 from carboxyl groups.
- **Transaminases** — Catalyse the transfer of amino groups.
- **Ligases** — Catalyse the synthesis of new C-O, C-S, C-C or C-N bonds, joining together two molecules.

Factors affecting enzyme activity

There are a great many factors which affect the activity of enzymes.

- They are **denatured** by **high temperatures** and extremes of **pH**, both of which alter the structure of the active site, preventing the substrate from binding with the enzyme.
- The rate of an enzyme-controlled reaction is affected by the **concentration of the substrate**. The concentration of substrate which is needed to make the reaction proceed at half its maximum rate is known as the **Michaelis constant**. This constant varies from one enzyme-controlled reaction to another but does give an indication of the affinity of the enzyme for its substrate. When the Michaelis constant is low, the reaction proceeds rapidly as the enzyme and substrate have a high affinity for one another.
- The presence of **inhibitors** can also have an effect on enzyme reactions, causing them to slow down or stop altogether. Inhibitors may be:
  - **competitive**, where the inhibitor molecule has a similar shape to the normal substrate molecule and competes for the active site of the enzyme; or
  - **non-competitive**, where the inhibitor molecule can either block the active site or cause an alteration in the shape of the enzyme molecule by attaching at some other site (e.g. it may attach to prosthetic group).
- Most enzymes can work in either direction and as metabolic reactions are **reversible**, the direction in which the reaction proceeds depends on the relative concentrations of substrate and product.
- In some metabolic pathways, the end-product of the pathway may act as an inhibitor. This is known as **end-product inhibition**, and it is an example of a **negative feedback mechanism**, preventing the unnecessary accumulation of a metabolite. A good example is shown in aerobic respiration, where the accumulation of ATP inhibits one of the respiratory enzymes in the pathway.

PRACTICAL WORK

There is a great deal of practical work which can be done in connection with the contents of this chapter. Every student of A-level Biology is expected to have a basic understanding of the nature of the biologically important molecules, and to be able to carry out simple tests to identify certain carbohydrates, lipids and proteins. It is recommended that the following tests be known:

> Make sure you are familiar with these tests

- for **starch** – the iodine test
- for **reducing sugars** – Benedict's test

    (It is not necessary to be able to identify specific reducing sugars, but the Benedict's test can be a **quantitative** test and it is not unusual to find examination questions which use it in this way.)

- for **non-reducing sugars** – boil with acid to hydrolyse, neutralise, then add Benedict's reagent
- for **proteins** – either the Biuret test or Millon's reagent (take care – poisonous!)
- for **lipids** – either stain with Sudan III (lipids take up the red stain) or shake up lipid with ethanol, add an equal volume of water; the resulting cloudy white precipitate indicates lipid. Neither of these tests for lipids is very satisfactory or suitable for all types of lipid material, and it is often better to rub the material thought to contain lipids on a piece of greaseproof paper; a translucent stain which persists after drying will indicate lipid.

Most syllabuses will expect candidates to have studied a range of different cell types showing how structure is related to function. This can be done from sections of tissues, prepared slides, squashes, smears, films or macerates. It is usually necessary to learn how to make temporary slide preparations of plant and animal material, and how to use simple staining techniques in making these temporary preparations. Some of these techniques may be tested in a practical examination, or form part of teacher-assessed practical work. The syllabus should be checked for the cells and tissues required, and whether or not a temporary preparation may be needed in a practical test.

A study of the fine structure of cells and organelles can be made from electron micrographs. These are photographs of material observed using the **electron microscope**. The sections of material have to be very thin, and are mounted on special grids. They are treated with special stains involving heavy metals and these scatter the electrons in the beam so that structures can be seen. This type of electron microscope is called the **transmission electron microscope (TEM)**, to distinguish it from the **scanning electron microscope (SEM)**, in which solid specimens can be viewed and surface features show up particularly well.

The characteristic features – the endoplasmic reticulum, ribosomes, Golgi apparatus, chloroplasts, mitochondria, nuclei, lysosomes, microtubules and microvilli – should be known, and reference will be made to these structures in the appropriate chapters.

There is a great deal of practical work which can involve enzymes, and it is wise to carry out and become familiar with several experiments involving different enzymes. This type of experiment is popular in practical examinations, and can be a very straightforward exercise if the principles are well understood. Syllabuses should be checked for any specific requirements. It is usual for enzymes such as amylase, sucrase and catalase to be named, and for experiments which demonstrate the effects of varying temperature, pH and enzyme concentration to be specified.

Further reading
Edelman J and Chapman J M, 1978, *Basic Biochemistry*. Heinemann
Kramer L M J and Scott J K, 1978, *The Cell Concept*. Macmillan
Pickering W R and Wood E J, 1982, *Introducing Biochemistry*. John Murray
Marshall D, 1983, *Cells and the Origin of Life*. ABAL. Cambridge University Press

EXAMINATION QUESTIONS AND ANSWERS

QUESTION, STUDENT ANSWER AND EXAMINER COMMENTS

Question 4.1

$$R - \underset{\underset{NH_2}{|}}{\overset{\overset{H}{|}}{C}} - COOH$$

Diagram for question 4.1

CHAPTER 4 EXAMINATION QUESTIONS

a) What general type of molecule is shown in the diagram above?

66 Correct answer 99 → amino acid (1)

b) What is the simplest form of **R**?

→ hydrogen atom – H (1)

c) which part of the structure gives acidic properties to the molecule?

66 Both correct → – COOH (carboxyl group) (1)
answers 99
d) Which part of the structure gives basic properties to the molecule?

→ – NH$_2$ (amino group) (1)

e) It may be said that because molecules of this type can show polymerisation they are very important biologically.
  i) What is meant by *polymerisation*?

66 Good answer but could The addition of large numbers of similar simple molecules (monomers) to form a giant
have incorporated the idea of
joining of molecules 99 molecule. (1)

  ii) If molecules of this type polymerise, what will be formed?

66 Could have put → protein (1)
polypeptide as an
alternative 99 f) With the aid of a diagram, illustrate the product when two of the units shown have joined

66 Correct way of setting
out the molecule 99

$$NH_2 - \underset{\underset{H}{|}}{\overset{\overset{R}{|}}{C}} - \underset{\underset{O}{\|}}{\overset{\overset{H}{|}}{C}} - N - \underset{\underset{H}{|}}{\overset{\overset{R}{|}}{C}} - COOH$$

(2)

g) What general type of biochemical reaction is this?

→ condensation (1)

66 Correct answer 99 h) What name is given to the type of bond formed between the two units?

→ peptide link (1)

(Total 10 marks)
(ODLE)

QUESTIONS

Question 4.2
Hydrogen peroxide reacts under certain conditions to give rise to free oxygen. The addition of manganese dioxide powder to a 3% solution of hydrogen peroxide in a test tube is followed by the production of a gas which causes a glowing splint to ignite. Explain the reaction.

Devise an experiment using fresh liver to investigate whether it also contains an ingredient which reacts in a similar way with hydrogen peroxide. A liquidiser is provided for use if required. What would you expect the active component to be? How would you test for this? What controls would you use?

Write out the experiment as you would in your own practical notebook. State the results you would expect from each test and give your interpretation of them.
(NICCEA Specimen Paper)

Question 4.3
a) Make a large labelled diagram of a generalised non-dividing plant cell. Underline the names of those organelles which can be seen under the light microscope. (5)
b) Distinguish between the structure and function of the **two** types of endoplasmic reticulum present in cells. (3)
c) Describe how DNA in the nucleus controls protein synthesis in the cytoplasm. (Do not include details of DNA replication.) (12)

(Total 20 marks)
(OCSEB)

OUTLINE ANSWERS

Answer 4.2

An explanation of the effects of manganese dioxide on hydrogen peroxide would involve discussion of the breakdown of the hydrogen peroxide into oxygen and water. Oxygen gas is identified by the ignition of a glowing splint. The manganese dioxide is acting as a catalyst: it is speeding up the rate of reaction.

In the experiment with the fresh liver, you would expect the active ingredient to be an enzyme, catalase, which is a biological catalyst. You could test for this by doing your experiment with ground fresh liver and with ground boiled liver. This way you can show that heating the enzyme will denature it and the reaction will not proceed. You would need to think carefully about the controls – hydrogen peroxide on its own, liver on its own. In devising the experiment, care should be taken to specify quantities of material used, whether ground up or in cubes and the temperature at which the reaction was carried out.

In writing up the experiment, be careful to describe your method accurately, including all the relevant details. You should include details of actual volumes of solutions used and actual quantities of material. Your explanation of the results should contain reference to the properties of enzymes and to the type of reaction being investigated.

Answer 4.3

A candidate would be expected to include some of the following aspects in answer to this question.

a) Marks are usually awarded for the quality of the diagram, its size, neatness and proportions. In this case, it would be appropriate to include as many features as you can, whether seen under the light or electron microscope.
b) The two types of endoplasmic reticulum are smooth and rough. The smooth is involved with the formation and transport of lipids and the rough, which is encrusted with ribosomes, is where protein synthesis takes place.
c) The description would need to include references to the information coded for by DNA, its transcription into messenger RNA and translation to the ribosomes in the cytoplasm. A detailed account, illustrated where appropriate, is needed, together with all the relevant information about transfer RNA in order to score high marks.

REVIEW SHEETS

1. Define each of the following, giving examples.
   a) Cellular level of organisation: _____
   _____
   b) Colonial organisation: _____
   _____
   c) Tissue level of organisation: _____
   _____

2. Define each of the following, giving examples.
   a) Kingdom: _____
   _____
   b) Phylum: _____
   _____
   c) Class: _____
   _____

3. What do you understand by the use of these terms?
   a) Genus: _____
   b) Species: _____
   c) Binomial system: _____
   d) Keys: _____

4. Identify the kingdom to which each of the following definitions relate.
   a) The members of this kingdom are eukaryotic, the body consisting of a network of threads called hyphae, forming a mycelium. _____
   b) The members of this kingdom are multicellular and have heterotrophic nutrition. ___
   c) The members of this kingdom are multicellular and photosynthetic. _____
   d) The members of this kingdom are **bacteria**, either non-photosynthetic, such as *Escherichia coli*, or photosynthetic, such as *Nostoc*. _____

5. Give examples of each of the following **classes**. Then identify the *phylum* and *kingdom* to which they belong. _____
   a) Class Monocotyledons _____
   _____
   b) Class Hydrozoa _____
   _____
   c) Class Arachnida _____
   _____
   d) Class Aves _____
   _____
   e) Class Gastropoda _____
   _____

6. Define each of the following, using examples and formulae where appropriate.
   a) Monosaccharides: _____
   b) Disaccharides: _____

CHAPTER 4 CELL BIOLOGY

c) Polysaccharides: _____

7 Outline some of the key properties of the **water molecule** which have some biological significance.
_____
_____
_____
_____

8 Draw a diagram to show the structure of an amino acid.

9 Describe some of the main properties and structures found in proteins.
_____
_____
_____
_____

10 Draw a diagram to show the structure of DNA. Briefly describe the structure and role of DNA.
_____
_____
_____
_____
_____
_____

11 Complete the labels for the following diagrams.

A typical plant cell   A generalised animal cell

12 a) Define enzymes and briefly describe some of their uses.
_____
_____

b) List some of the factors affecting the activity of enzymes.
_____
_____

CHAPTER 5
AUTOTROPHIC NUTRITION

- HOLOPHYTIC NUTRITION
- PHOTOSYNTHESIS
- REQUIREMENTS FOR PHOTOSYNTHESIS
- LIGHT-DEPENDENT REACTIONS PHOTOPHOSPHORYLATION
- LIGHT-INDEPENDENT REACTIONS
- C_3 AND C_4 PLANTS
- PRACTICAL WORK

GETTING STARTED

Autotrophic nutrition involves the **synthesis of complex food molecules** so that an understanding of the nature and composition of carbohydrates, fats and proteins is helpful in understanding the pathways involved. In addition it is necessary to appreciate how **amino acids** are formed and then built up into **proteins**. It is also important to note that most syllabuses do not require biochemical details of individual reactions in the metabolic pathways involved, so that it is not necessary to remember complex chemical formulae.

A general understanding of the structure of the flowering plant is necessary, together with a knowledge of leaf anatomy, particularly in relation to the exchange of gases. The structure of **chlorenchyma tissue**, or a **palisade cell**, is a syllabus requirement for some boards, and the structure of a **chloroplast** as revealed by the electron microscope is common to all syllabuses.

When considering the different stages of the process of **photosynthesis**, it could be helpful to refer frequently to a diagram of the fine structure of the chloroplast, so that each stage can be related to its correct site.

CHAPTER 5 AUTOTROPHIC NUTRITION

ESSENTIAL PRINCIPLES

HOLOPHYTIC NUTRITION

Autotrophic nutrition involves the synthesis of organic compounds from inorganic raw materials and it is often referred to as holophytic nutrition, which literally means 'feeding like plants'. There are two types of autotrophic nutrition: **photosynthesis** and **chemosynthesis**.

PHOTOSYNTHESIS

> Photosynthesis and Chemosynthesis are types of autotrophic nutrition

This is the process by which green plants, algae and certain types of bacteria build up **complex organic molecules** from **carbon dioxide**, **water** and **mineral ions**. The source of **energy** for this process comes from **light**, which is absorbed by **chlorophyll** and related pigments. These organisms are sometimes called **photo-autotrophs**: i.e. they use **inorganic** sources of carbon and light energy for their synthetic processes.

CHEMOSYNTHESIS

This is the process by which a few bacteria can perform a similar synthesis of organic compounds using energy derived from special methods of respiration. These organisms are sometimes referred to as **chemo-autotrophs**: i.e. they use inorganic sources of carbon and **chemical** sources of energy for their synthetic processes.

This chapter deals exclusively with the process of photosynthesis in green plants and algae. It is worth noting that animals can feed only on **complex** organic molecules, which they must obtain from plants and other animals. This is known as **heterotrophic** nutrition, and these two types of nutrition pinpoint the basic differences between green plants and animals.

PHOTOSYNTHESIS

The whole process is often summarised by a general equation:

$$CO_2 + H_2O \xrightarrow[\text{chlorophyll}]{\text{light}} (CH_2O)n + O_2$$

> General equation for photosynthesis

$$\text{carbon dioxide} + \text{water} \xrightarrow[\text{chlorophyll}]{\text{light}} \text{carbohydrate} + \text{oxygen}$$

From this it can be deduced that the plant must take in carbon dioxide, water and light energy in order to synthesise carbohydrate, oxygen gas being produced as a waste product. It should be remembered that other syntheses follow from the availability of simple organic molecules and energy; amino acids are built up for the production of proteins, and fatty acids and glycerol, which are involved in the formation of lipid molecules.

REQUIREMENTS FOR PHOTOSYNTHESIS

CARBON DIOXIDE

Carbon dioxide is obtained from the atmosphere by **diffusion** into the leaf via **stomata** down a concentration gradient. The concentration of carbon dioxide in the atmosphere is about 0.03% and on a warm, sunny day when the **light intensity** and **temperature** are both high it can be the factor which **limits** the rate of photosynthesis.

The necessity for carbon dioxide can be demonstrated by enclosing a leaf of a well-watered, **destarched** potted plant, such as a geranium, in a flask containing potassium hydroxide. The potassium hydroxide will absorb carbon dioxide from the air in the flask. A similar leaf on the same plant should be enclosed in a control flask without the potassium hydroxide. After several hours in suitable light conditions, both leaves are detached and tested for the presence of starch using iodine solution. Starch should be

> Carbon dioxide can limit the rate of photosynthesis

present in the control leaf, showing that photosynthesis has occurred, but not in the leaf deprived of carbon dioxide. The plant is destarched initially by keeping it in the dark for 48 hours so that any starch present originally has been converted to soluble carbohydrate and transported away from the leaves or used for respiration. This enables a valid conclusion to be drawn from the results of the experiment.

WATER

Water is obtained from the soil via the **roots** and is transported to the leaves through the **xylem tissues** of the root, stem and petioles to the lamina of the leaf. Water is necessary for a great many metabolic reactions taking place in the plant, and it is possible to trace the fate of the water used in the photosynthetic process by supplying a plant with water containing the heavy isotope of oxygen $H_2^{18}O$.

LIGHT

Light is essential for photosynthesis. This can be demonstrated easily by keeping a potted plant in darkness for 48 hours and testing a leaf with iodine solution to show that starch is not present. Then part of one of the leaves, still attached to the plant, is covered up and the whole plant is exposed to suitable light conditions for several hours. When the leaf is detached and tested for starch, only those parts which were exposed to light, i.e. uncovered, should give a positive result, a blue-black colour, indicating that photosynthesis has occurred. The covered parts of the leaf will remain the colour of the iodine solution, yellow-brown, showing that no starch has been synthesised.

Intensity of light

> The intensity of light affects the rate of photosynthesis

The intensity of the light can be an important factor in determining the rate of photosynthesis, and small changes can be quite significant, both on a daily basis and also with respect to habitat. Increasing the light intensity causes the rate of photosynthesis to increase, up to a critical point, at which some other factor becomes limiting. Above this point, any increase in the light intensity will have no further effect on the rate. The relationship between light intensity and rate is a linear one at low levels, as shown in the graph in Fig. 5.1.

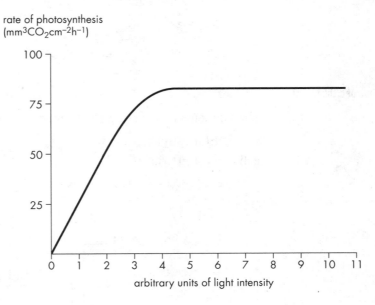

Fig. 5.1 Graph to show relationship between light intensity and the rate of photosynthesis

Wavelength of light

> The wavelength of light is an important factor

The wavelength of light is also important, and it can be shown that chlorophyll absorbs only certain wavelengths, reflecting others. This is done by passing a beam of light through an extract of chlorophyll and then through a prism which will separate out the different wavelengths. If after passing through the prism the light is projected on to a screen, it can be seen that the **red** and the **blue** parts of the **spectrum** disappear, leaving the green in the middle. This is an **absorption spectrum** and only indicates which wavelengths have been absorbed by the photosynthetic pigments; it does not indicate whether the wavelengths are actually used. This can be shown by plotting an **action spectrum** of the amount of carbohydrate synthesised by plants exposed to different wavelengths of light. Fig. 5.2 shows the results and also indicates that the absorption spectrum and the action spectrum for the chloroplast pigments show a close correlation.

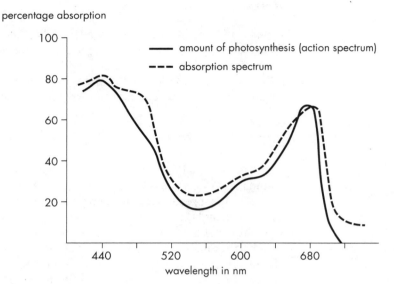

Fig. 5.2 The action spectrum for photosynthesis and the absorption spectrum for chlorophyll plotted on the same graph

PHOTOSYNTHETIC PIGMENTS

The role of the **photosynthetic pigments** is to absorb light energy and to convert it into chemical energy. The pigments are located on the **chloroplast membranes**. This arrangement, together with the arrangement of the chloroplasts within the cells, allows for the maximum absorption of the available light (see Fig. 5.3).

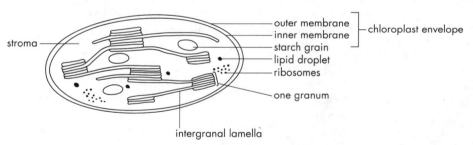

Fig. 5.3 Structure of a chloroplast

The pigments in flowering plants are of two types:

- the **chlorophylls**, a and b
- the **carotenoids**

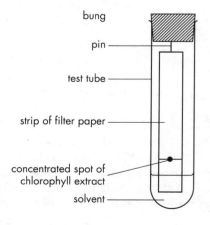
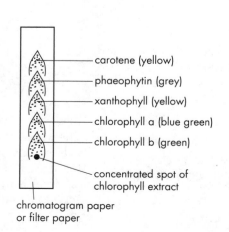

Fig. 5.4 Paper chromatogram of photosynthetic pigments from a green plant

> The role of the pigments in flowering plants

Chlorophylls

The **chlorophylls** absorb mainly in the red and blue-violet regions of the spectrum (see Fig. 5.4). **Chlorophyll a** is a bluish-green pigment which is common to all photosynthetic organisms, playing a central role in the absorption of light. **Chlorophyll b** is yellowish-green and is found in the higher plants and green algae. Both types are **porphyrin** molecules, containing a head with a central **magnesium** ion and a long **hydrocarbon** tail.

Carotenoids

The **carotenoids** are yellow-orange pigments and they absorb **blue-violet** light. Their presence is masked by the green chlorophylls and it is thought that, in addition to acting as **accessory pigments**, they help to protect the chlorophylls from excessive light. The most widespread of the carotenoids is β-**carotene**, an orange pigment found in all photosynthesising plants.

Xanthophylls

Xanthophylls are similar in structure to carotenes, but are yellow.

LEAF STRUCTURE AND FUNCTION

At this point, before considering the biochemistry of phosynthesis, it is worth considering how well the structure of the leaves of flowering plants is related to their function. The lamina is thin so that diffusion paths for gases are short; the epidermis is transparent, allowing light penetration; the chloroplasts are abundant in the palisade cells, enabling maximum light absorption; the spongy mesophyll tissue allows for the circulation of gases; the stomata permit gas exchange; and the vascular tissues provide support and transport (see Fig. 5.5.).

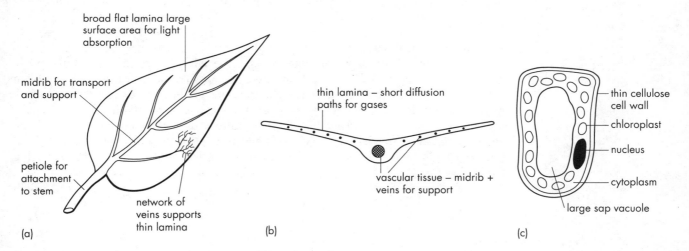

Fig. 5.5 Adaptations of a green leaf for photosynthesis a) whole leaf- external view b) section through lamina c) single palisade cell

PHOTOSYNTHETIC PROCESS

Photosynthesis is a two-stage process, involving initially the formation of **ATP (adenosine triphosphate)** and the **reduction of NADP (nicotinamide adenine dinucleotide phosphate)** during the **light-dependent reactions**, followed by a **light-independent stage** in which carbon dioxide is reduced in the synthesis of organic compounds.

LIGHT-DEPENDENT REACTIONS – PHOTOPHOS-PHORYLATION

These reactions involve the **photochemical** splitting of water, yielding hydrogen for the reduction of fixed carbon dioxide in the second stage, coupled with the formation of ATP molecules by **photophosphorylation**. These reactions take place in the **grana** of the chloroplasts.

It has been mentioned already that it is possible to trace the fate of the water used in photosynthesis with the use of the heavy isotope of oxygen to label the water. The following balanced equation shows that all the oxygen evolved in photosynthesis is derived from the water:

$$CO_2 + 2H_2{}^{18}O \Rightarrow (CH_2O) + {}^{18}O_2 + H_2O$$

"An important balanced equation"

This equation shows that water is produced as well as used in the process of photosynthesis, and it is a much more accurate summary of events than the equation given earlier.

When light energy is absorbed by the photosynthetic pigments, they become **excited** and emit **electrons**, which are accepted by other molecules called **electron acceptors**. This is an **oxidation-reduction** process in which the pigment molecule, the **electron donor**, is **oxidised** and the acceptor molecule is **reduced**.

There are two functional types of pigments: **primary pigments** and **accessory pigments**. The latter pass on the electrons they emit to the primary pigments. The electrons emitted by the primary pigments are used to drive the reactions of the photosynthetic process.

Current opinion follows the theory that two types of photosynthetic unit exist, called **Photosystem I (PSI)** and **Photosystem II (PSII)**. each photosystem contains a collection of accessory pigment molecules passing energy on to one primary pigment molecule (reaction centre). From observations made, it has been estimated that each photosystem consists of sub-units; each sub-unit contains about 300 chlorophyll molecules. Chlorophyll b is an accessory pigment molecule and passes on its energy to chlorophyll a, which is a primary pigment molecule. PSI is at a higher energy level than PSII.

CYCLIC AND NON-CYCLIC PHOTOPHOSPHORYLATION

Photophosphorylation is the conversion of ADP to ATP involving light energy and inorganic phosphate ions.

Cyclic photophosphorylation

In **cyclic** photophosphorylation, light causes the energy level of an electron to be raised and the electron is emitted from the chlorophyll molecule. It is taken up by an electron acceptor molecule and passed along a chain of electron carriers, which include **cytochromes**, at different energy levels. As the electron is passed from one carrier to the next, energy is removed and used in the synthesis of ATP molecules. Eventually the electron achieves its normal energy level and returns to the chlorophyll molecule. In this type of photophosphorylation only PSI is involved and only ATP is formed (see Fig. 5.6).

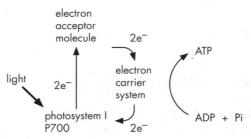

Fig. 5.6 Cyclic photophosphorylation

Non-cyclic photophosphorylation

In **non-cyclic** photophosphorylation, light initiates the emission of electrons from PSI and PSII and electron acceptors are reduced. PSI and PSII are both oxidised. PSII is restored to electrical neutrality by electrons from the splitting of water. **Oxygen** is produced as a waste product. PSI is neutralised by electrons emitted from PSII which have lost some of their energy in the formation of ATP. Some of the electrons combine with **hydrogen ions**, from the splitting of water, forming **hydrogen atoms**, which are then taken up by **NADP** to give **NADPH + H⁺** (see Fig. 5.7). Non-cyclic photophosphorylation can be summarised in the following equation, where P represents inorganic phosphate:

$$H_2O + NADP + 2ADP + 2P \Rightarrow \tfrac{1}{2}O_2 + 2(NADPH + H^+) + 2ATP$$

LIGHT-INDEPENDENT REACTIONS

These reactions take place in the **stroma** of the chloroplast and they do not require light. As they are controlled by **enzymes**, the rate at which they take place is affected by temperature. The reactions use ATP as a source of energy and NADPH + H$^+$ as the source of the reducing power. Both of these are produced in the light-dependent stages.

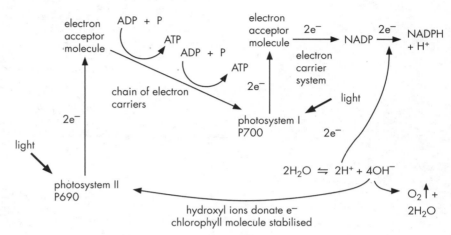

Fig. 5.7 Non-cyclic photophosphorylation

EVENTS IN THE CALVIN CYCLE

The sequence of events in this stage of photosynthesis was worked out by Calvin and his associates, using ^{14}C, a radioactive isotope of carbon, and the unicellular green alga, *Chlorella*. His experiments yielded evidence for the role of a **five-carbon** acceptor molecule, **ribulose bisphosphate (RuBP)**, which takes up the carbon dioxide, forming an unstable six-carbon compound. This six-carbon compound immediately splits into two molecules of a **three-carbon** compound called **phosphoglyceric acid (PGA)**. PGA is **reduced** by means of the products of the light stages, the energy needed coming from the hydrolysis of ATP. **P**hosphoglyceraldehyde (PGAL) is formed and some of this three-carbon sugar can be built up into **glucose phosphate** and then into **starch** by **condensation**. In order that the cycle continues, the majority of the PGAL formed is used to regenerate molecules of RuBP, the acceptor molecule (see Fig. 5.8).

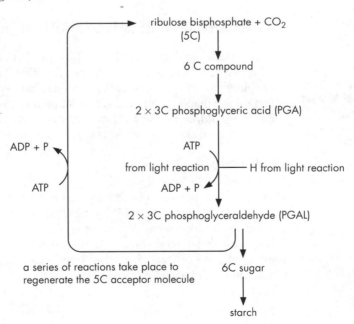

Fig. 5.8 Events in the Calvin cycle

Products of photosynthesis

Carbohydrates are not the only products of the photosynthetic process, as was shown clearly in Calvin's experiments, where the radioactive carbon was found to be incorporated into lipid and amino acid molecules at an early stage. Fig. 5.9 shows an outline of the pathways involved.

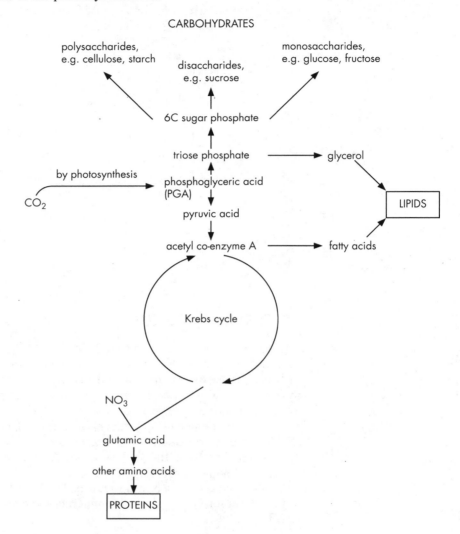

Fig. 5.9 An outline of the synthetic pathways involved in the production of lipids and amino acids

FACTORS AFFECTING THE RATE OF PHOTOSYNTHESIS

Factors such as **light, carbon dioxide** and **temperature** have already been mentioned as having an effect on the rate of photosynthesis. Students should appreciate that in many habitats the interrelation of these factors will have a significant effect on the plants, both on a daily and on a seasonal basis. The **principle of limiting factors** should be understood: i.e. that the rate of a reaction will be governed by the factor that is nearest its lowest relative concentration.

> Be familiar with the 'principle of limiting factors'

Experimental work

It is possible to investigate the factors controlling photosynthesis with a series of experiments using a pondweed such as *Elodea*, which will give off bubbles of oxygen if it is photosynthesising rapidly enough. The bubbles can be counted quite easily, or the gas can be collected and the volume measured using the Audus apparatus (see Fig. 5.10). This relatively simple method can be used to investigate the effects on photosynthesis of varying the light intensity, carbon dioxide concentrations, temperature and wavelengths of light.

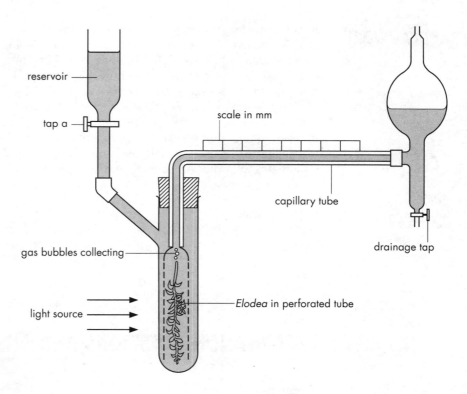

Fig. 5.10 Audus apparatus

C₃ AND C₄ PLANTS

The events described above, where the first products of photosynthesis are three-carbon compounds, are typical of the alga *Chlorella* and most of the plants found in temperate regions. They are now referred to as **C_3 plants**.

In the 1960s it was found that in some tropical plants, such as maize and sugar cane, the first compounds to appear labelled with ^{14}C were 4-carbon organic acids. Hatch and Slack worked out the metabolic pathway for these plants, designating them **C_4 plants**. It was shown that the carbon dioxide acceptor molecule was a 3-carbon compound, **phosphoenolpyruvate (PEP)**.

Fixation of the carbon dioxide was very rapid and occurred in the cytoplasm of the mesophyll cells; four-carbon acids were formed and then moved into special cells around the vascular bundles. The carbon dioxide was then released by decarboxylation of the organic acids and fixed via the usual pathway using RuBP and the Calvin Cycle.

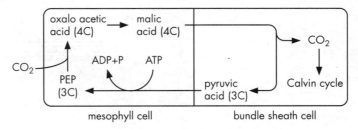

Fig. 5.11 Outline of C_4 fixation

PEP = phospho enol pyruvic acid
3C acceptor molecule

The advantages of this type of initial fixation by PEP are that it is rapid (PEP carboxylase has a higher affinity for CO_2 than RuBP carboxylase) and that a high concentration of carbon dioxide is built up in the bundle sheath cells. This increases the efficiency of photosynthesis and avoids the process of **photorespiration** taking place when carbon dioxide concentrations are low.

- **Photorespiration** occurs when the concentration of oxygen is high and the concentration of carbon dioxide is low. The enzyme which catalyses the uptake of carbon dioxide by RuBP, **RuBP carboxylase**, will accept oxygen as a substrate as well as carbon dioxide, so that the two gases compete for the **active site** of the enzyme. Oxygen is a **competitive inhibitor** of carbon dioxide fixation. Increasing the carbon dioxide concentration will favour photosynthesis, but a decrease will favour photorespiration.

CHAPTER 5 AUTOTROPHIC NUTRITION

PRACTICAL WORK Suggestions for practical work have been made where relevant in the various sections of this chapter. The more simple experiments, demonstrating the need for carbon dioxide, chlorophyll and light, will have been encountered already, but determinations of the rate of photosynthesis and how it is affected by the various factors may be less well known. These experiments can be set up quite easily and yield good results.

The relevance of photosynthesis to agriculture and horticulture is important and the application of the principles to the improvement of yields is highly relevant. The results of even quite simple class experiments can emphasise the significance of the controlling of environmental conditions in glasshouses in order to improve yields. Candidates following AS courses may find options in Plant Biology where this topic is studied in more detail.

Further reading
Hall D O and Rao K K, 1983, *Photosynthesis*. Arnold
Pickering W R and Wood E J, 1982, *Introducing Biochemistry*. John Murray

EXAMINATION QUESTIONS AND ANSWERS

Question 5.1
Write an account of the dark reaction (light-independent reactions) in photosynthesis and explain how radioactive tracers have been used to study this pathway. *(20)*
(ULEAC)

Question 5.2
The apparatus shown below is used for collecting and measuring the volume of oxygen given off by the pondweed *Elodea* during photosynthesis.

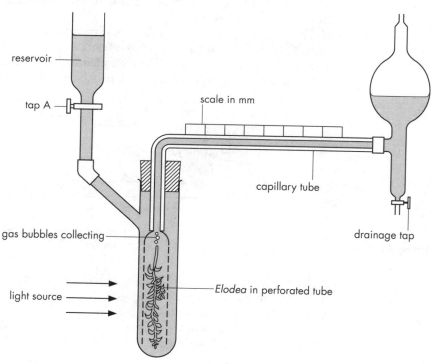

a) How would the following be maintained during the experiment?
  i) Constant temperature
  ii) Constant carbon dioxide supply to the pondweed. *(2)*
b) For what purpose would tap A be opened during the experiment? *(1)*
c) Why is capillary tubing used in the apparatus? *(2)*

d) How would you use this apparatus to obtain an accurate measurement of the rate of photosynthesis of the *Elodea*? (4)
e) How could you determine the effect of wavelength of light on the rate of photosynthesis using the same apparatus? (3)

(Total 12 marks)
(ULEAC)

Question 5.3

An investigation was carried out into the effect of carbon dioxide concentration and light on the productivity of lettuces in a glasshouse. The productivity was determined by measuring the rate of carbon dioxide fixation in milligrams per dm^2 leaf area per hour.

Experiments were conducted at three different light intensities, 0.05, 0.25 and 0.45 (arbitrary units), the highest approximating to full sunlight. A constant temperature of 22°C was maintained throughout.

The results are given in the table below.

CARBON DIOXIDE CONCENTRATION/PPM	PRODUCTIVITY AT DIFFERENT LIGHT INTENSITIES/MG DM^{-2}/H^{-1}		
	At 0.05 units light intensity	At 0.25 units light intensity	At 0.45 units light intensity
300	12	25	27
500	14	30	36
700	15	35	42
900	15	37	46
1100	15	37	47
1300	12	31	46

a) For the experiment at 0.25 units of light intensity, describe and comment on the effect on the productivity of the lettuces of increasing carbon dioxide concentration in the following ranges:
   i) 300 to 900 ppm (2)
   ii) 900 to 1300 ppm (2)
b) i) A carbon dioxide concentration of 300 ppm is approximately equivalent to that in atmospheric air.
   For each of the three light intensities, work out the maximum increase in productivity that was obtained compared with that at 300 ppm and use it to calculate the percentage increase in productivity at each light intensity.
   1 At 0.05 units light intensity
   2 At 0.25 units light intensity
   3 At 0.45 units light intensity (3)
   ii) Comment on the effect on productivity of changing light intensity. (2)
c) Explain why the carbon dioxide concentration affects the productivity of plants. (3)
d) State why the temperature should be kept constant during this experiment. (1)
e) Suggest why, even with artificial lighting, glasshouse crops generally need to have more carbon dioxide added when temperatures are low, than when temperatures are high. (2)

(Total 15 marks)
(ULEAC)

OUTLINE ANSWERS

Answer 5.1

Candidates must ensure that they spend enough time on both parts of this question in order to score a good mark. Diagrams can be used in both parts but it is unlikely that maximum marks can be gained using diagrams only.

In the first part, credit would be given for reference to the pathway of the carbon dioxide from the atmosphere to the chloroplast stroma, where it is fixed. Carbon dioxide is then picked up by the ribulose bisphosphate, under the influence of the carboxylase enzyme, RuBP carboxylase. A 6-carbon compound is formed which is unstable and which breaks down immediately into two molecules of 3-carbon phosphoglyceric acid (PGA). The PGA is phosphorylated and then reduced to give phosphoglyceraldehyde (PGAL), using ATP and NADPH from the light reactions. Most of the PGAL is used to regenerate the carbon dioxide acceptor molecule, RuBP. This involves a complex series of reactions. PGAL can be converted to hexose sugars and then to starch by condensation. Other organic compounds, amino acids and fatty acids are also produced.

Calvin used radioactive tracers in his experiments with the green alga *Chlorella*, trying to trace the sequence of products formed in photosynthesis. He used the radioactive isotope of carbon, ^{14}C. This was incorporated into hydrogencarbonate and supplied to the algae in a specially designed 'lollipop' apparatus. The algae were illuminated and allowed to photosynthesise for specified times and then samples were removed and analysed. The compounds containing the radioactive carbon were identified. By allowing the experiments to proceed for different periods of time, it was possible to trace the sequence of products.

Credit would be given for any correct details of the procedure and for any other experiments.

Answer 5.2

a) i) Immerse in a water bath.
   ii) Hydrogencarbonate could be dissolved in the water.
b) The tap would be opened to clear bubbles through or to move the bubbles to the scale.
c) Capillary tubing is used so that accurate measurements of the small volume of gas can be made.
d) The apparatus would need to be equilibrated with the temperature of the water bath, and then all the bubbles cleared by opening the tap. The apparatus should be allowed to run for a specified time, then the tap used to move any gas to the scale, where the volume can be read and recorded. For greater accuracy, several readings should be taken and a mean worked out. The rate can be worked out by dividing the volume by the time.
e) The apparatus should be kept in constant light intensity and at a constant temperature, and then exposed to different coloured lights or a lamp covered with different coloured filters. The volume of gas evolved at each wavelength should be recorded.

Answer 5.3

a) i) A sharp increase from 300 to 700 ppm, as more CO_2 is available for fixation. From 700 to 900 ppm, rate of increase is less as other factors are limiting.
   ii) From 900 to 1,000 ppm the productivity stays the same as CO_2 is not limiting. From 1,100 to 1,300 ppm the productivity falls as excess CO_2 inhibits productivity.
b) i) 1 Max increase   = 3 units
         % increase   = 3/12 × 100 = 25%
      2 Max increase   = 12 units
         % increase   = 12/25 × 100 = 48%
      3 Max increase   = 20 units
         % increase   = 20/27 × 100 = 74%
   ii) Increased light intensity increases the productivity as there is a greater energy input.
c) At high light intensity, CO_2 limits productivity because the atmospheric concentration is low. Carbon dioxide in the air provides the carbon for forming organic compounds.
d) This is to eliminate temperature as a variable.
e) The rate of the dark reaction is affected by temperature as it is enzyme controlled. At low temperatures, increased CO_2 concentrations may increase the reaction rate.

CHAPTER 5 ESSENTIAL PRINCIPLES

QUESTION, STUDENT ANSWER AND EXAMINER COMMENTS

Question 5.4

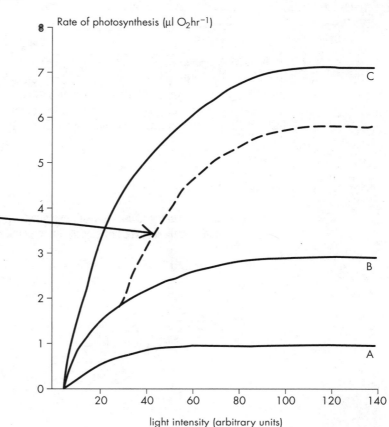

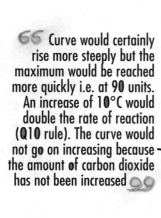

❝ Curve would certainly rise more steeply but the maximum would be reached more quickly i.e. at 90 units. An increase of 10°C would double the rate of reaction (Q10 rule). The curve would not go on increasing because the amount of carbon dioxide has not been increased ❞

a) The graphs illustrate the results obtained in experiments performed to find out what effects increased carbon dioxide levels could have on the rate of photosynthesis in one type of annual crop plant. All experiments were carried out in controlled conditions at 16°C. The amount of artificial light given to the plants could be varied.

Rate of photosynthesis (as measured by evolution of oxygen) plotted against light intensity for three groups of plants supplied with different amounts of carbon dioxide.

Group A: with 0.01% (by vol.) carbon dioxide.
Group B: with 0.03% (by vol.) carbon dioxide.
Group C: with 0.15% (by vol.) carbon dioxide.

i) At what light intensity do those plants given 0.03% CO_2 reach their maximum rate of photosynthesis under these conditions?

    90 units   (1)

❝ Correct answer read off from graph B ❞

ii) Suggest a reason for the form of graph A beyond 54 units of light.

    No increase in rate due to lack of CO_2 becoming a limiting factor   (1)

❝ Graph A reaches maximum and there is then no increase; reasoning correct ❞

iii) Suggest a reason for the form of graph C beyond 54 units of light.

    Still enough CO_2 for increasing rate of photosynthesis with light intensity. Doesn't become a limiting factor until later.   (1)

❝ Correct answer ❞

iv) Which graph best indicates the rate of photosynthesis one might expect to obtain from this species of plant in air at 16°C?

    Group B   (1)

❝ Correct; the amount of carbon dioxide in the atmosphere is 0.03% ❞

v) Suggest why the graphs do not commence at the origin.

    Below a certain light intensity (5 units) there is not enough energy for photosynthesis to begin.   (1)

❝ Even at 5 units there is no photosynthesis; answer given is adequate ❞

CHAPTER 5 AUTOTROPHIC NUTRITION

vi) State **two** possible ways (other than amount of evolved oxygen) by which the rate of photosynthesis of the plants might have been measured.

1) By the amount of a product such as starch built up over a specific length of time.

2) By the amount of $^{14}CO_2$ assimilated in a specific length of time. (2)

> 66 Answer given as 1. is fine, but it would be difficult to do 2. Perhaps increase in dry mass would be better 99

vii) Suppose that, with Group B plants, the temperature had been increased by 10°C at 30 light units and above. Draw on the graph, with a dotted line, any difference you might expect to the given curve.

b) If the light intensity is reduced to a very low level the plants will be below the compensation point. Explain what is meant by this term.

Compensation point is when the net gas exchange equals zero i.e. In gas exchange, the rate of photosynthesis exactly balances the rate of respiration (2)

(Total 11 marks)
(ODLE)

> 66 Correct answer; could have been explained by saying that the amount of carbon dioxide used in photosynthesis exactly balances the amount of carbon dioxide released by respiration 99

NB: You can find Review Questions on this Chapter at the end of Chapter 6.

CHAPTER 6 HETEROTROPHIC NUTRITION

- TYPES OF HETEROTROPHIC NUTRITION
- FEEDING MECHANISMS
- DIET
- THE ALIMENTARY CANAL
- THE ROLE OF THE LIVER
- PRACTICAL WORK

GETTING STARTED

Heterotrophic nutrition is the acquisition of food materials by organisms in the form of complex organic molecules, and a knowledge of the nature of these organic molecules is helpful in the full understanding of this topic.

The structure of **enzymes**, the way in which they function and the factors affecting their activity are fundamental to an understanding of the processes of **digestion**. These topics have already been covered in Chapter 4 (Cell Biology).

It would also be helpful to have worked through Chapter 5, Autotrophic Nutrition, in order to appreciate the interdependence of the green plants and the heterotrophic organisms. Green plants are the **producers**, the major group of organisms utilising energy from the sun in the synthesis of food materials upon which all heterotrophic organisms are directly or indirectly dependent.

CHAPTER 6 HETEROTROPHIC NUTRITION

TYPES OF HETEROTROPHIC NUTRITION

❝ Types of heterotrophic nutrition ❞

ESSENTIAL PRINCIPLES

A heterotrophic organism cannot synthesise its major food requirements and so is dependent on a source of complex organic molecules. The food is required as a source of energy for activities such as **locomotion**, and for the **synthesis of body tissues** in the processes of **growth**, **repair** and **replacement**.

There are three major types of heterotrophic nutrition:

- holozoic
- parasitic
- saprotrophic (saprophytic)

HOLOZOIC NUTRITION

Holozoic nutrition literally means 'feeding like animals'; this involves taking in **complex** organic molecules, breaking them down by **digestion, absorption** into the body tissues from the digestive system and finally **utilisation** of the absorbed products of digestion in the body cells. Animals that feed solely on plant material are termed **herbivores**, those that feed on other animals are **carnivores**, and the **omnivores** have a mixed diet.

PARASITIC NUTRITION

A **parasite** is an organism that lives in or on another living organism, its **host**. The parasite derives all its nutrition from the host, together with a certain amount of shelter in many cases. The host does not gain any benefit from the association and is often harmed in some degree. At the very least it is deprived of a certain amount of food. The most successful parasites are those that do least harm to their hosts, thus ensuring their own survival. Some parasites live on the outside of their host, the **ectoparasites**: e.g. fleas, leeches and the flowering plant called *Cuscuta*, or dodder. Others live entirely within the body of the host and are termed **endoparasites**: e.g. *Plasmodium*, the malarial parasite, and *Taenia*, the tapeworm. Parasites are considered to be very highly specialised organisms and show considerable adaptations to their mode of life. The adaptations are concerned with enabling the parasite to remain attached to its host; to gain its nutrition in the most efficient manner; and then to produce as many offspring as possible in order to gain entry into a new host organism.

SAPROTROPHIC NUTRITION

Saprotrophic organisms feed on dead, decaying organic matter. They secrete enzymes on the food substrate and then absorb the products of this **extracellular digestion**. The dead organic matter could be made up of both plant and animal parts and the activities of these organisms are important in the decomposition of litter and the recycling of valuable nutrients.

Many syllabuses refer to parasites and saprotrophic organisms, so it would be sensible to check which examples are relevant. Reference is often made to a mould fungus such as *Mucor* or *Rhizopus*, and details of this type of nutrition are represented in Fig. 6.1.

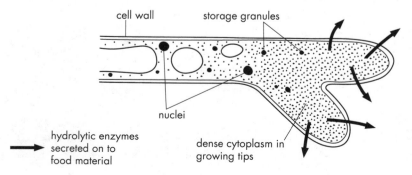

Fig. 6.1 Mucor hypha showing extra-cellular digestion

CHAPTER 6 ESSENTIAL PRINCIPLES

FEEDING MECHANISMS

A heterotrophic organism has to obtain and take in its food before it can begin the process of digestion. This may involve specialised mechanisms or **mouthparts**, depending on the nature of the food. There are very many different structures involved, and one way of looking at the topic is to group the animals according to the type of food material taken in.

> *Useful groupings of animals*

- Animals that take in small particles are **microphagous**, e.g. the earthworm.
- Animals that take in large particles are **macrophagous**, e.g. the sheep and the dog.
- Animals that take in liquid food are **fluid feeders**, e.g. aphids and butterflies.

MICROPHAGOUS FEEDERS

The **microphagous** feeders ingest small particles by means of **pseudopodia** or by **ciliary currents**. In *Amoeba*, the pseudopodia surround and enclose particles of food together with a drop of water. This forms a **food vacuole** inside which digestion occurs. The soluble products of digestion then diffuse into the cytoplasm, leaving the undigested material, which is egested. In *Paramecium*, the beating of **cilia** in the **oral groove** wafts food particles towards the **cytopharynx**, where **food vacuoles** are formed. (See Fig. 6.2) In *Mytilus* (the common mussel), the gills are covered with cilia which beat, creating a current of water which enters through the **inhalant siphon**. The microscopic food particles are trapped in **mucus** and then swept into the mouth. The food is sorted out and water leaves the animal via the **exhalant siphon**.

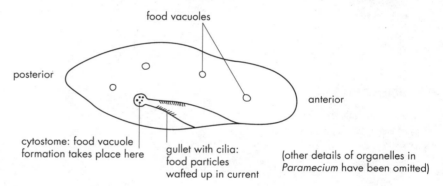

Fig 6.2 Paramecium showing food vacuole formation

Mytilus is an example, relevant to some syllabuses, of the many **sedentary** aquatic animals that set up feeding currents which waft food particles into their bodies where digestion takes place.

MACROPHAGOUS FEEDERS

There is a great variety of feeding methods found amongst the **macrophagous** feeders, from **tentacles** in *Hydra*, a **radula** in the snail and biting and chewing mouthparts in insects to the true **teeth** found in the vertebrates. Very few syllabuses ask for details of all these feeding methods, but many do require a knowledge of teeth in mammals, some even specifying herbivorous, carnivorous and omnivorous dentition.

FLUID FEEDERS

The **fluid feeders** have sucking mouthparts in the form of a tubular structure. Most of the fluid feeders are insects and the tubular structures are formed either from a modified **labium**, as in the case of the housefly, or from the **maxillae**, as in the butterfly. Mosquitoes and aphids have piercing structures in addition, as they need to penetrate through the tissues of their host in order to obtain food.

TEETH

The teeth of mammals break up large masses of food in the mouth, a process often referred to as **mechanical digestion**, making the food easier to swallow as well as increasing the surface area available for subsequent enzyme action. Mammals are **heterodont**, having different types of teeth adapted for different functions. There are three functionally adapted types of teeth:

> Types of teeth

- the **incisors**, at the front of the mouth, for cutting and biting off chunks of food
- the **canines**, at the sides, for gripping and holding prey
- the **cheek** teeth: **premolars** and **molars**, at the sides and back of the mouth, for crushing, grinding, chewing and slicing food

The canines are usually well pronounced in carnivores, such as the cat and dog, but poorly developed in humans and absent or reduced in many herbivores.

The cheek teeth are usually highly specialised to deal with a specific type of food: **grinding** premolars and molars in the sheep and other herbivores, for dealing with vegetation (see Fig. 6.3b); sharp **carnassial** teeth in the carnivores for slicing through flesh (see Fig. 6.3a).

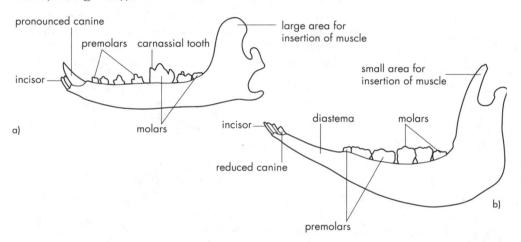

Fig. 6.3 a) Carnivore lower jaw bone; and b) Herbivore lower jaw bone

An examination of the skulls of a herbivore, a carnivore and an omnivore would be useful in establishing the differences in dentition related to diet. It is also interesting to relate diet to jaw articulation.

No syllabus requires a knowledge of the development of teeth in humans, but the generalised structure of a tooth and reference to dental disease are requirements in some.

DIET

All heterotrophs must obtain sufficient **energy-providing** and **growth-promoting** food in their diet to satisfy their needs.

BALANCED DIET

Balanced diets must include the correct amounts and proportions of carbohydrates, proteins and lipids, together with specified **vitamins** and **mineral ions**. Human diets vary according to the individual's age, sex, activity and body size. It is important that the protein included in the human diet should provide all the amino acids essential for the tissues – a factor to be considered by vegetarians, as protein from some plant sources does not include all these amino acids.

Vitamins and mineral ions are usually required only in small amounts and contribute to the formation of essential coenzymes and cofactors in many metabolic reactions. It is only too easy to remember them by the symptoms produced if there is a deficiency in the diet, but emphasis should be placed on their roles in metabolism. Tables of vitamins and mineral ions essential in the human diet are to be found in the standard texts to which reference should be made for the details you need for your syllabus.

THE ALIMENTARY CANAL

DIGESTION

Mouth

Food enters the **buccal cavity**, where it is broken up by the teeth in mechanical digestion. While chewing is in progress, the food is also mixed with **saliva** from the **salivary glands**. There are three pairs of these glands and they are stimulated to release saliva by the sight and smell of food, as well as by its presence in the mouth.

Saliva is a watery secretion, containing **mucus** and **salivary amylase**, together with some mineral ions. It is important for lubricating the food before it is swallowed. Salivary amylase is an enzyme which begins the digestion of any cooked starch in the ingested food to **maltose**.

The food is formed into a **bolus**, moved to the back of the mouth by the tongue and then swallowed. It passes into the **oesophagus** or **gullet**, where it is moved down to the **stomach** by **peristaltic** contractions of the longitudinal and circular muscles.

Swallowing is a **reflex** action, triggered off by the food touching the **soft palate**. The **nasal cavity** is closed and the **epiglottis** closes over the **glottis**, preventing the food from entering the trachea.

Stomach

The presence of food in the mouth stimulates the secretion of the **gastric juice** from the **gastric glands** in the stomach wall, so that when food enters the stomach digestion may begin.

The **stomach** is a J-shaped storage sac for food. It has a very muscular wall which contracts rhythmically and mixes up the food with the **gastric juice**. The gastric glands are simple tubular glands with **peptic cells**, **oxyntic cells** and **mucus-secreting cells** at the top.

- **Peptic cells** secrete the enzymes as inactive precursors, **pepsinogen** and **pro-rennin**; they are protein-digesting enzymes and could damage the stomach tissues before being released; when they are activated, they are prevented from damaging the stomach wall by the secretions of mucus.
- **Oxyntic cells** secrete **hydrochloric acid**, which makes the stomach contents acid; the acid conditions kill off many pathogenic bacteria as well as activating the protein-digesting enzymes; **pepsinogen** becomes **pepsin** and begins the breakdown of proteins to **polypeptides**; **pro-rennin** becomes **rennin**, which coagulates the soluble protein in milk, so this enzyme is especially important in young mammals.
- The supply of gastric juice from the gastric glands is maintained by the hormone **gastrin**, secreted from the stomach wall into the blood.

The food is churned up by the muscular action until it is in a semi-liquid mass called **chyme**. This is emptied into the **duodenum** by periodic relaxation of the **pyloric sphincter** (see Figs 6.4a) and b)).

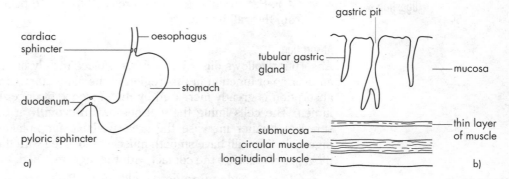

Fig. 6.4a) Human stomach; and b) Section through stomach wall

Small intestine

The small intestine in humans is divided into two main regions, the **duodenum** and the **ileum**.

- The duodenum comprises the first 20 cm of the small intestine and receives secretions from both the **liver** and the **pancreas**.

- **Bile** is produced in the liver and stored in the **gall bladder**, from where it passes into the duodenum via the **bile duct**. It contains no enzymes, but it does contain bile salts, **sodium** and **potassium glycocholate** and **taurocholate**, which are important in emulsifying the lipids present in the food. **Emulsification** is achieved by lowering the surface tension of the lipids, causing large globules to break up into tiny droplets. This enables the action of the enzyme **lipase** to be more efficient as the lipid droplets now have a much larger surface area. Bile also contains **bile pigments**, formed as the result of the breakdown of the **haemoglobin** from damaged and worn-out red blood cells. In addition, there is **sodium hydrogencarbonate**, which helps to neutralise the acidity of the food as it comes from the stomach.
- The **pancreatic juice** is secreted from the **exocrine glands** in the **pancreas**. This juice contains a number of different enzymes:
  - **pancreatic amylase**, which breaks down any remaining **starch** to **maltose**
  - **pancreatic lipase**, which splits **lipids** into **fatty acids** and **glycerol**
  - **trypsin**, secreted as inactive **trypsinogen**, which continues the breakdown of proteins
  - **nucleases**, which break down **nucleic acids** to **nucleotides**

  The pancreatic juice enters the duodenum through the **pancreatic duct**.

In the walls of the duodenum and the ileum are **Brunner's glands**, which secrete an **alkaline** juice and **mucus**. The alkaline juice helps to keep the contents of the small intestine at the correct pH for enzyme action, and the mucus is for lubrication and protection.

Enzymes are secreted from groups of cells at the bottom of the **crypts of Lieberkuhn**. These enzymes include:

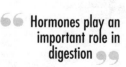
Enzymes in the pancreatic juice

- **enterokinase**, which activates **trypsinogen** to **trypsin**
- **maltase**, which splits **maltose** into two **glucose** residues
- **sucrase**, which splits **sucrose** into **glucose** and **fructose**
- **endopeptidases**, and **exopeptidases**, which complete the digestion of **proteins** to **amino acids**

At the completion of digestion, the carbohydrates are in the form of monosaccharides, the proteins as amino acids and the lipids either as fatty acids and glycerol or as very tiny droplets.

The secretion of the digestive juices in the small intestine is controlled by **hormones**. The presence of acid food in the duodenum stimulates the production of two hormones from the wall of the duodenum: **secretin** and **cholecystokinin-pancreozymin** (CCK-PZ).

Hormones play an important role in digestion

- **Secretin** causes the liver to secrete bile and the pancreas to produce the fluid part of the pancreatic juice.
- **CCK-PZ** stimulates enzyme release from the pancreas and the release of bile from the gall bladder.

Absorption

Absorption follows digestion and takes place mainly in the small intestine, so there is an overlap of function in this region of the alimentary canal. The surface available for absorption is greatly increased by the presence of finger-like **villi** projecting from the lumen. The cells lining the villi possess **microvilli** at their free surfaces and these serve to further increase the area available for uptake of the soluble products of digestion. The villi have smooth muscle associated with them, and this enables them to move, so increasing the contact with the digested food.

- **Glucose** and **amino acid** molecules These are absorbed across the epithelium of the villi by a combination of diffusion and active transport; they pass into the capillary network which supplies each villus and are then transported away to the **hepatic portal vein** and thence to the liver.
- **Fatty acid** and **glycerol** molecules These pass into the epithelial cells and recombine to form **neutral fat**, which is then passed into the **lymphatic system**; eventually the fats pass into the bloodstream via the **thoracic duct**.

- **Vitamins** and **mineral salts** These do not undergo digestion and are absorbed from the small intestine.
- **Water** Absorbed from the colon.
- **Food residues** These consist of undigested material, cellulose, bacteria and any inorganic substances which are in excess; they move relatively slowly through the colon, becoming more solid as water is removed; when they reach the **rectum** they are in a semi-solid state and are egested via the **anus** as **faeces**.

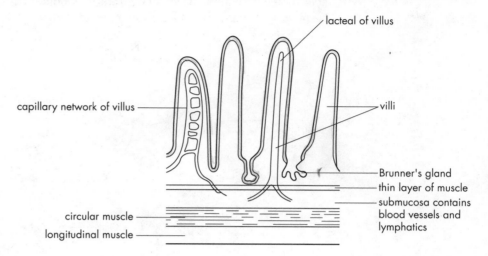

Fig. 6.5 Section through ileum wall

VARIATIONS IN ALIMENTARY CANALS

The alimentary canals of mammals vary slightly depending on the diet. In the **ruminants**, such as sheep and cows, there are a number of **oesophageal pouches** into which the food passes when it has been cropped. Here there are colonies of **cellulose-digesting bacteria** which mix with the food and break down the plant cell walls. The food is **regurgitated** and re-chewed (chewing the cud), and when it is re-swallowed it bypasses these chambers and enters the true stomach, where digestion proceeds as in other mammals (see Fig. 6.6). Fatty acids are formed from the bacterial action, and these are used by the ruminant.

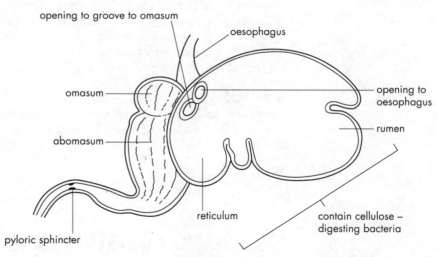

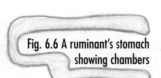

Fig. 6.6 A ruminant's stomach showing chambers

In other herbivores, the cellulose-digesting bacteria are found in a much enlarged **caecum** and **appendix**.

ROLE OF THE LIVER

Blood glucose regulation

The soluble products of digestion are carried to the liver in the **hepatic portal vein**. If the **blood glucose** level is high, then the liver can remove glucose from the blood and convert it to **glycogen** for storage.

Lipid and amino acid regulation

The liver also plays a part in the regulation of **lipids** and **amino acids.** Lipids are either broken down or modified for storage in **adipose tissue**. Excess amino acids cannot be stored in the body and must be broken down by **deamination**. The amino group is removed from each amino acid, converted first to **ammonia** and then to **urea** by a cycle of events in the liver cells called the **ornithine** cycle. The urea is shed into the blood and eventually removed by the kidneys, while the remainder of the amino acid molecule enters the carbohydrate metabolic pathways and is either oxidised in respiration or converted to glycogen and stored.

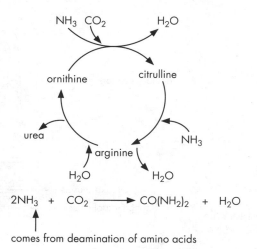

Fig. 6.7 Ornithine cycle

PRACTICAL WORK

The practical work associated with this topic ranges from the observation of the life history and modes of nutrition of the heterotrophic organisms, through enzyme experiments linked to the hydrolytic gut enzymes and on to an understanding of the structure of teeth and the alimentary canal of a small mammal.

Enzyme experiments have been described elsewhere, but some experiments can be done using amylase, maltase and sucrase, together with lipase and protein-digesting enzymes. These exercises will help in the understanding of the processes of digestion.

Some syllabuses require candidates to dissect the alimentary canal of a small mammal as part of the teacher-assessed practical work. Whether or not a dissection is mandatory, it is sensible to familiarise oneself with the structure of the digestive system and its associated structures, such as the liver and the pancreas. In addition some syllabuses require a knowledge of the basic anatomy of the main regions of the gut, and the ways in which these are adapted to their functions.

Further reading

Dissection guides for the anatomy of the mammalian gut.
Rowett H C Q, *Guide to Dissection*. John Murray
Freeman W H and Bracegirdle B, 1967, *Atlas of Histology*. Heinemann

EXAMINATION QUESTIONS AND ANSWERS

QUESTIONS

Question 6.1
a) Distinguish between autotrophic and heterotrophic nutrition. (5)
b) Describe the processes of digestion and absorption of lipids in a mammal. (10)
c) Explain why lipids are suitable storage compounds in living organisms. (5)

(Total 20 marks)
(ULEAC)

Question 6.2
a) Heterotrophs include animals and microorganisms. Compare their methods of feeding.
b) What is a balanced diet?
c) What are the essential components of a balanced diet and what are:
   i) their functions in metabolism;
   ii) the effects of both their excess and their deficiency?

(NICCEA Specimen Question)

Question 6.3
The table below refers to the alimentary canals of a sheep and a dog.
  If the statement is correct for that animal place a tick (✓) in the appropriate box and if the statement is incorrect place a cross (✗).

STATEMENT	SHEEP	DOG
Crowns of teeth entirely covered with enamel		
Diastema pesent		
Stomach region of adult expanded into pouches containing symbiotic microorganisms		
Rennin present in gastric juice in young		
Alimentary canal connected by mesentery to body wall		

Question 6.4
The diagram below shows a transverse section through the stem of a parasitic angiosperm dodder (*Cuscuta*) and part of a host plant.

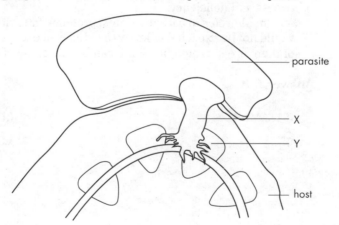

a) i) Identify the structure labelled X and state its function. (2)
   ii) Identify the host tissue labelled Y, with which structure X is in contact. (1)
b) State **two** ways in which extensive parasitism by dodder may damage the host plant. (2)
c) Parasites commonly show a reduction or loss of some of the features possessed by closely related organisms that are not parasitic.
   i) State **three** such features reduced or lost in adult dodder plants. (3)
   ii) Briefly explain how such reduction or loss may have arisen in parasitic organisms. (2)

(Total 10 marks)
(ULEAC)

CHAPTER 6 HETEROTROPHIC NUTRITION

OUTLINE ANSWERS

Answer 6.1

a) Autotrophic organisms use inorganic sources of carbon; heterotrophic organisms must be supplied with organic carbon.
   Autotrophic organisms synthesise complex organic molecules using light energy or chemical energy; examples are green plants, nitrogen cycle bacteria.
   Heterotrophic organisms break down complex organic compounds to simple ones by digestion using enzymes; examples are animals, fungi, bacteria.

b) Reference should be made to the following points: mechanical digestion by teeth in the mouth; mechanical churning and warming in the stomach; fat globules pass into duodenum; meet with bile from the gall bladder; release of bile stimulated by CCK-PZ; bile salts named; emulsion formed, increasing the surface area of the fat globules for the action of the enzyme lipase from the pancreatic juice; lipase splits the fats into fatty acids and glycerol by hydrolysis; pancreatic juice released under the influence of secretin; the products diffuse into the epithelium of the villi of the ileum; some fats pass into the lacteals of the lymphatic system.

c) Fats have a high calorific value and are very good energy stores; they are compact molecules so do not take up much room; they are non-osmotic so do not affect cell water relations; they give protection, buoyancy and insulation; on oxidation metabolic water is formed; credit given for any suitable examples of the above.

Answer 6.2

Brief outline answer:

a) Nutrition of animal (digestion inside the organism) compared with bacterium or fungus (enzymes secreted on to food, soluble products absorbed).

b) Balanced diet definition: the correct components in the correct proportions.

c) Components list; need function in metabolism: i.e. what is it there for and effects of excess and deficiency.
   Should refer to carbohydrates, proteins, fats, vitamins and minerals. Candidates would not be expected to know the effects of excess of the individual vitamins: some of them are excreted but some can be poisonous if taken in excess.

Answer 6.3

STATEMENT	SHEEP	DOG
Crowns of teeth etc	✗	✓
Diastema present	✓	✗
Stomach region pouches etc	✓	✗
Rennin present etc	✓	✓
Alimentary canal etc	✓	✓

Answer 6.4

a) i) Haustorium for attachment, to obtain nutrients from the host.
   ii) Phloem.

b) — deprives the host of nutrients.
   — may allow the entry of pathogens.

c) i) Reduced leaves; lack of chlorophyll; loss of roots.
   ii) Non-essential parts use up resources; some mutants would arise with small parts; these mutants would be favoured by selection.

CHAPTER 6 ESSENTIAL PRINCIPLES

QUESTION, STUDENT ANSWER AND EXAMINER COMMENTS

Question 6.5
The diagram below shows a longitudinal section of part of the ileum wall.

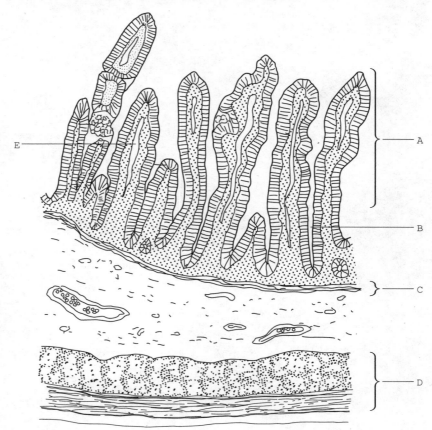

a) Name structures A, B and C.

A Villus — *Correct*

B Crypt of Lieberkühn — *Correct*

C Mucosa — *Incorrect, should be muscularis mucosae or muscle layer*

(3)

b) Explain how structures A, D and E enable the ileum to function efficiently.

A This gives a large surface area for the absorption of digested products. — *Worth 2 marks*

D This makes peristalsis possible. — *Insufficient, should also give a comment on the antagonistic action of circular and longitudinal muscles and the role of peristalsis*

E This is for the absorption of fat breakdown products. — *Insufficient, should also mention transport from the gut via the lymphatic system*

(6)

(Total 9 marks)

REVIEW SHEET

1 Using an appropriate general equation, outline the process of photosynthesis.

$CO_2 + H_2^{18}O \rightarrow CH_2O + {}^{18}O_2$

2 Consider the role of each of the following in the process of photosynthesis.
   a) Carbon dioxide: _____
   b) Water: _____
   c) Light: _____

3 Discuss the role of the *chlorophylls* and *carotenoids* as photosynthetic pigments.

4 Explain the meaning of the following balanced equation

$CO_2 + 2H_2^{18}O \Rightarrow (CH_2O) + {}^{18}O_2 + H_2O$

5 Draw a diagram and use it to describe **cyclic photophosphorylation**.

6 Describe the main events in the **Calvin Cycle**.

7 Compare and contrast C_3 with C_4 plants.

8 Compare **autotrophic nutrition** with **heterotrophic nutrition**.

9 Describe each of the following.
   a) Holozoic nutrition:

   b) Parasitic nutrition:

   c) Saprotrophic nutrition:

10 Describe, with examples, each of the following.
   a) Microphagous feeders:

   b) Macrophagous feeders:

   c) Fluid feeders:

10 Consider the role of the **stomach** in digestion.

11 With the aid of a diagram, explain how **absorption** occurs.

12 Briefly outline the role of the **liver** in heterotrophic nutrition.

CHAPTER 7
HOMEOSTASIS

MAMMALIAN KIDNEY AS A HOMEOSTATIC ORGAN

MAMMALIAN LIVER AS A HOMEOSTATIC ORGAN

THERMOREGULATION

HYPOTHALAMUS AND PITUITARY GLAND

WATER BALANCE IN LOWER ANIMALS

PRACTICAL WORK

GETTING STARTED

The term **homeostasis** refers to the maintenance of the stability of the internal environment within the body, giving an organism a level of internal independence from the external environment. Most physiological processes are either involved in homeostasis or are dependent upon its existence. Homeostatic mechanisms are important in all living organisms, but in A level most emphasis is placed on homeostatic mechanisms in mammals. This chapter will deal with aspects of homeostasis in mammals which are not found elsewhere in the book, but cross-reference should be made to the control of respiration and gas exchange, Chapter 8, and the regulation of blood glucose and energy release, Chapter 10. The **mammalian kidney** is central to homeostasis, being involved in the regulation of water and salt balance, pH and metabolite levels. The **liver** is centrally concerned in the regulation of metabolite levels and **thermogenesis**. The **hypothalamus** monitors pH, osmotic pressure and temperature. It brings about regulation either by direct effect on the nervous system or via the hormones from the **pituitary gland**. **Negative feedback** mechanisms play an important role in keeping metabolite levels and physiological functions within the narrow limits necessary for continued existence. These are dealt with in Chapter 10.

CHAPTER 7 HOMEOSTASIS

ESSENTIAL PRINCIPLES

Organisms live in a constantly changing external environment. Cells live in an internal environment which must be kept constant if these cells are to survive and function correctly. The homeostatic mechanisms within an organism are designed to keep the internal environment constant. By their nature, homeostatic mechanisms involve the **interaction** of many organs and consequently an understanding of this process involves your synthesising your knowledge of the functions of many organs and systems within an organism.

MAMMALIAN KIDNEY AS A HOMEOSTATIC ORGAN

> The excretory function is only part of the kidney's role in homeostasis

FUNCTION OF THE KIDNEY

- Removal of **nitrogenous metabolic waste** from the body. This is a form of homeostatic function as wastes interfere with metabolic activity if they are allowed to accumulate.
- Maintenance of the **fluid composition** of the body. This must be maintained within narrow limits even though the intake of water, food and salts varies widely. Water and salts are also lost through the skin and water is lost via the lungs; the kidneys must compensate for the losses. All the materials to be excreted by the kidneys are dissolved in the blood plasma, as are valuable materials such as glucose and amino acids. The kidney has to carry out its excretory functions whilst **conserving** these necessary chemicals.

COMPOSITION OF KIDNEY

A kidney is composed of millions of **nephrons**, each made up of four functional parts: **Bowman's capsule**, a **proximal tubule**, the **loop of Henle** and a **distal tubule** leading to a **collecting duct** (see Fig. 7.1).

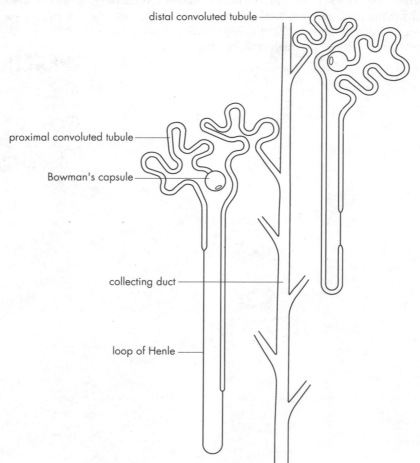

Fig. 7.1 Kidney tubules

NEPHRON FUNCTION OF KIDNEY

Three different processes – ultrafiltration, selective reabsorption and secretion – are involved in nephron function.

Ultrafiltration

Ultrafiltration is the process by which small molecules such as water, glucose, urea and salts are filtered from the knot of capillaries, the **glomerulus**, into **Bowman's capsule**. Most of the pressure producing the filtration comes from the hydrostatic pressure of the blood in the glomerular capillaries. This pressure is opposed by the pressure in the capsule produced by the narrow efferent vessels and also by the water potential in the blood produced by the colloidal plasma proteins. These pressures are shown in Table 7.1.

The glomerular pressure can be altered by changes in the diameter of the **afferent** and **efferent** arterioles entering and leaving the glomerulus.

	PRESSURE (kPa)
Pressure in glomerular capillaries	9.2
Pressure in Bowman's capsule	2.0
Resulting net filtration pressure	7.2
Water potential in blood	–4.0
Resulting net pressure available	3.2

Table 7.1 The forces involved in glomerular filtration

SELECTIVE REABSORPTION

Selective reabsorption of all the substances useful to the body takes place in different parts of the tubule.

Reabsorption of water/dissolved solutes

Water and dissolved solutes are absorbed at equal rates in the proximal tubule. Absorption is mainly by **passive** means. The surface area for this purpose is increased by the **brush border** of **microvilli** on the inner surface of the cells. The presence of many mitochondria indicates that there may also be active uptake (see Fig. 7.2).

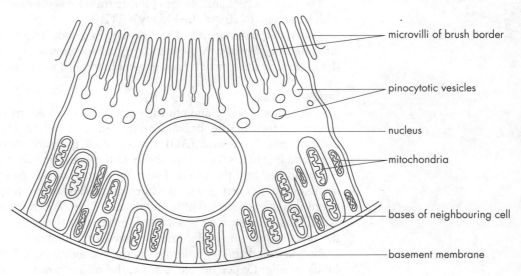

Fig. 7.2 Proximal tubule cell (as seen by electron microscopy)

Reabsorption of chloride ions

> Water content is not regulated by the loop of Henle

The loop of Henle is concerned with the reabsorption of chloride ions. Sodium ions follow these passively into the interstitial fluid of the medulla. The loop acts as a counter-current exchange system. This leads to an increase in the ionic concentration of the medulla. The descending limb of the loop is impermeable to solutes but permeable to water. The concentration of salts in the medulla thus leads to the osmotic loss of water from the descending limb. In the ascending limb the concentration of the fluid decreases again because of the movement of chloride (and sodium) out of the tubule. These movements are summarised in Fig. 7.3.

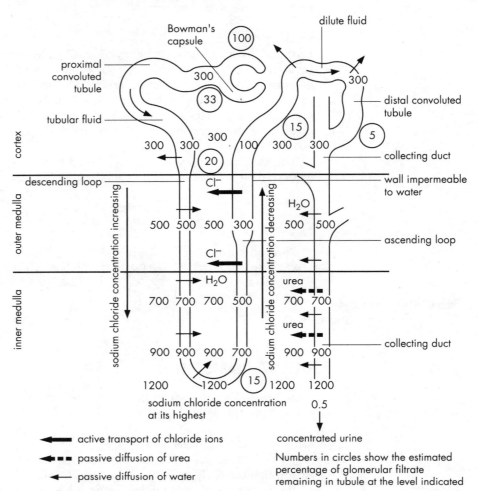

Fig. 7.3 Changes in concentration of tubular fluid as it passes through a kidney tubule

The process responsible for chloride transfer is the chloride pump, an active process which uses up to 70% of the kidney's ATP.

The water passing out of the descending loop enters the capillaries, **vasa recta**, surrounding the loop. Most selective reabsorption takes place in the distal tubule and the collecting duct, which are permeable to both water and urea.

Feedback system

The amounts of water and chloride reabsorbed are controlled by a feedback system. If osmoreceptors in the hypothalamus detect a fall in water potential of the blood, **antidiuretic hormone (ADH)** is secreted from the posterior pituitary gland. This hormone increases the permeability of the collecting duct to water and so more water is reabsorbed into the blood. The maintenance of plasma chloride (and sodium) levels in the blood is controlled by the hormone **aldosterone** from the adrenal cortex, which alters the salt reabsorption in the distal tubule. When the concentration of ions in the blood falls, aldosterone is secreted, increasing the active transport of chloride across the distal tubule. A similar effect occurs if the blood volume falls. Chloride is actively reabsorbed and this is followed by the osmotic uptake of water. Although the distal tubule is slightly permeable to urea, the majority of the urea is excreted in the urine, which also contains other nitrogenous wastes (uric acid, creatinine, ammonia) and any foreign materials filtered through the glomerulus (see Fig 7.4).

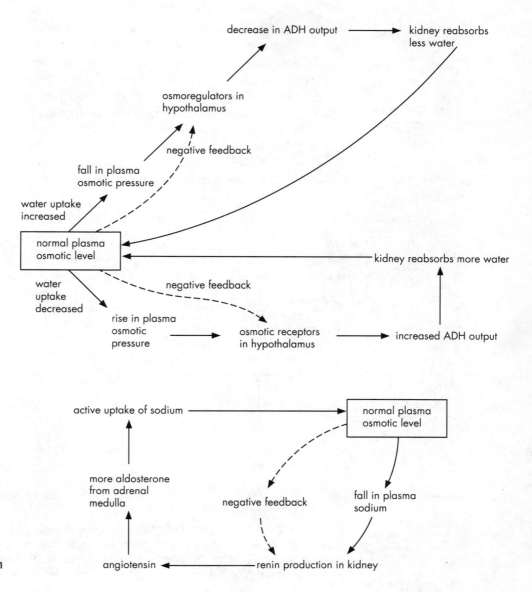

Fig. 7.4 A negative feedback system

SECRETION

> Do not forget secretion as a kidney function

Secretion of further substances not required by the body may take place in the distal convoluted tubule: e.g. hydrogen and hydrogencarbonate ions. This is important in the control of plasma pH, which must be maintained at pH 7.4. If the plasma pH falls, hydrogen ions are excreted by the kidney; if the plasma pH rises, hydrogencarbonate ions are secreted. A fall in pH also causes the kidney cells to produce ammonium ions, which combine with acids brought to the kidneys; the resulting ammonium salts are excreted.

MAMMALIAN LIVER AS A HOMEOSTATIC ORGAN

FUNCTION OF THE LIVER

The liver is central to the metabolism and is responsible for many homeostatic functions. At any given time, about 20% of the circulating blood is passing through the liver and, with the kidney, it is responsible for **maintaining metabolic levels in the blood**. Although the metabolic activity of the liver is complex, the structure is simple. The functional unit is an **acinus** as shown in Fig. 7.5. The homeostatic function of the liver is enhanced by the double blood supply. Blood comes from the heart via the **hepatic** artery; there is also a supply of venous blood rich in digested food via the **hepatic portal vein** from the gut.

- Maintenance of blood glucose levels in carbohydrate metabolism, with the aid of **insulin** and **glucagon** from the pancreas The liver is responsible for maintaining the blood glucose level at approximately 90mg 100cm^3 by the production of glycogen from glucose or vice versa; if the glycogen levels are low, the liver can produce more from fatty acids or amino acids residues.

- Removal of the amino groups from surplus amino acids The production of the excretory product **urea** from them (the ammonia formed from the amino group is converted to urea via the ornithine cycle – see Fig. 7.6).
- Removal of other unwanted materials such as the haemoglobin from old red blood cells, hormones which have served their purpose and toxic substances such as bacterial endotoxins, alcohol and drugs. These are all broken down by the liver.
- The production of hormone precursors (**steroids**), fatty acids, amino acids and plasma proteins are also aspects of the liver's homeostatic function.

> All liver functions are involved in homeostasis

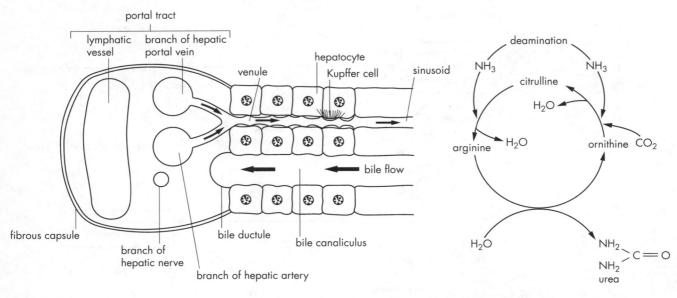

Fig. 7.5 A liver acinus

Fig. 7.6 Ornithine cycle

THERMO-REGULATION

> Do not use the term exothermic

The majority of animals are **ectothermic** ('cold blooded', poikilothermic): that is, their temperature varies with the external temperature. Birds and mammals are **endothermic** ('warm blooded', homoiothermic): they can maintain an internal temperature largely independent of the environmental temperature.

ECTOTHERMS

Terrestrial ectotherms have behavioural and physiological mechanisms which help them to gain or lose heat. These include colour changes, orienting the body to alter the amount of heat gained from the sun, avoiding direct sunlight, being able to alter the blood flow to the skin and using the evaporation of water for cooling.

ENDOTHERMS

Endothermic animals have a homeostatic mechanism controlled by the **hypothalamus**.

Heat gain
The major source of heat in endotherms is the exothermic metabolic activity within the cells, particularly the liver and the skeletal muscles. When the body is cold, the muscles start to contract voluntarily (**shivering**), producing additional heat. Some mammals also have special heat-producing tissue, **brown fat**, which metabolises rapidly under sympathetic stimulation.

Heat loss
Heat loss is controlled by altering the level of loss through the body surface. The rate of heat loss from the skin by conduction, convection and radiation depends on the blood flow through it. The skin of endothermic animals is highly vascularised, the rate

of flow through the skin capillaries being controlled by the **thermoregulatory centre** in the hypothalamus. The flow rate can be increased 100 fold in hot conditions. The production of **sweat** is also controlled by the hypothalamus. Sweating begins as soon as the body temperature rises above the normal value of 37°C. Very hairy mammals increase heat loss by licking the fur (heat being lost by evaporation), by panting or by loss through the nasal and buccal cavities.

HYPOTHALAMUS AND PITUITARY GLAND

The hypothalamus participates in many of the homeostatic processes. In some cases – for example, temperature regulation – the hypothalamus acts directly via the **autonomic nervous system**. In other cases the hypothalamus acts by producing hormones, **releasing factors**, which cause the pituitary gland to release **trophic hormones**. These in turn stimulate other endocrine glands. It also produces some **inhibiting factors**. The production of both releasing factors and trophic hormones is governed by negative feedback (see Fig. 7.7). The trophic hormones are released by the anterior pituitary. The posterior pituitary produces ADH, which is concerned in homeostasis.

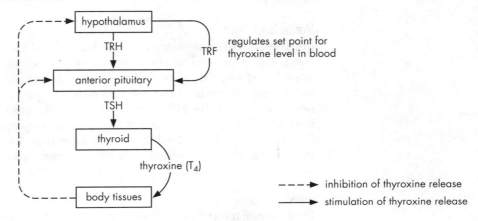

Fig. 7.7 Control system for thyroxine secretion

WATER BALANCE IN LOWER ANIMALS

The control of water and salt balance is of vital importance to all animals. The processes in some animals are outlined below.

FRESHWATER PROTOZOA

Freshwater protozoa have a more negative water potential than the water medium and so gain water by **endosmosis**. To counteract this they have **contractile vacuoles**. Water from the cytoplasm collects into small vesicles which fuse with the vacuole and empty into it (see Fig. 7.8). There are many mitochondria associated with a contractile vacuole and these supply the energy necessary to keep the fluid in the vacuole against an osmotic gradient.

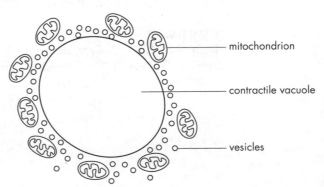

Fig. 7.8 Contractile vacuole and its vesicles

INSECTS

Insects living in dry conditions have two ways of reducing water loss.

- The presence of a waxy **epicuticle** covering the exoskeleton which prevents evaporation.
- Water loss by excretion is kept to the minimum by the **Malpighian tubules**, blindly ending tubules from the mid-gut lying in the haemocoel. Sodium and potassium urates are formed in the haemocoel and pass into the tubules. Here they react with carbon dioxide and water to form hydrogencarbonates and uric acid. The hydrogencarbonates and water pass back into the haemocoel and the uric acid crystallises out, to be lost from the body with minimum water.

BONY FISH

Bony fish (**teleosts**) have different problems depending on whether they are marine or freshwater.

Marine bony fish

Marine teleosts have body fluids which have a less negative water potential than sea water. The scales and mucus on the fish's surface make this impermeable to water but the gill surfaces gain salts and lose water. To maintain the balance, the fish drink sea water; secretory cells in the gut pass salts into the blood by active transport; **chloride secretory** cells in the gills actively pass chloride ions from the blood to the sea; sodium ions are also secreted to maintain the electrochemical balance; a small volume of isotonic urine is formed by the glomerular kidneys; the excretory product, **trimethylamine oxide**, and salts are secreted into the tubules and water follows by osmosis.

Freshwater bony fish

Freshwater teleosts have a more negative water potential than their environment has and consequently there is considerable endosmosis of water into the body and loss of ions through the permeable gill surface; they need to lose large volumes of water and to conserve salts. To maintain the balance they have highly glomerular kidneys, which produce a large volume of filtrate; salts are selectively reabsorbed through the kidney tubules, producing a large volume of dilute urine; the excretory waste is ammonia. This is present in the urine and is also lost through the gills; the chloride secretory cells in the gills actively take up ions from the surrounding water to help replace those lost.

PRACTICAL WORK

Various aspects of homeostasis can be investigated practically. The anatomical relationship of the excretory system, liver and endocrine glands to other structures can be investigated using models or by dissection. Experiments on the influence of surface/volume ratio on heat loss can be set up using flasks of boiling water. Similarly, the effect of insulation can be investigated. Damp paper towels or cotton-wool wadding can be used to simulate the evaporation of sweat.

Further reading

Hardy R N, 1983, *Homeostasis*, 2nd edn. Arnold

Monger G (ed), 1985 *Revised Nuffield Advanced Biology, Biology Study Guide 1* Longman.

CHAPTER 7 EXAMINATION QUESTIONS

EXAMINATION QUESTIONS AND ANSWERS

QUESTIONS

Question 7.1
For each section, ring the correct letter.
i) Besides eliminating waste nitrogenous materials the kidney also carries out the important function of osmoregulation. This is a process whereby the kidney:
  A selectively reabsorbs excess glucose under the influence of insulin
  B maintains the osmotic potential of the blood that enters it at the same osmotic potential as that leaving it
  C prevents excess water leaving the body in order to prevent dehydration
  D selectively accumulates useful mineral ions to maintain a constant osmotic potential of the blood
  E reabsorbs water in varying quantities in order to maintain a constant osmotic potential of the blood
ii) Which one of the following statements concerning kidney function is **incorrect**?
  A Glucose is normally reabsorbed in the first part of the tubule.
  B The kidneys require much energy for the process of reabsorption.
  C Water is reabsorbed in the proximal convoluted tubule.
  D The kidneys expend much energy in order that the process of ultra-filtration can occur.
  E Blood leaving the kidney contains less glucose and oxygen than the blood entering the kidney.
iii) Which one of the following statements concerning the liver's role in excretion is **incorrect**?
  A Excess amino acids are deaminated in the liver by removal of NH_2 and H.
  B In the urea (or orthinine) cycle, ammonia is effectively combined with carbon dioxide.
  C Deamination results in products which can be utilised in respiration.
  D Anti-diuretic hormone influences the rate of urea synthesis, exerting its controls through the enzyme arginase.
  E Three amino acids, arginine, orthinine and citrulline, are involved in the production of urea.

(ODLE)

Question 7.2
An investigation was carried out to determine the rate of flow and the composition of fluids in human kidneys. These were measured at positions A, B, C and D, shown in the diagram below, which represent a nephron and associated blood vessels. The results are given in the table below.

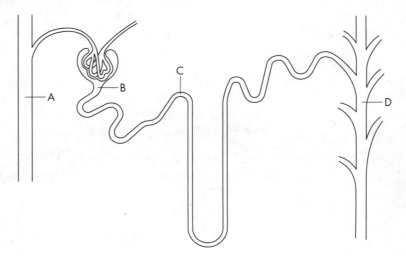

Fig. 7.9

CHAPTER 7 HOMEOSTASIS

POSITION	TOTAL FLOW RATE $cm^3 min^{-1}$	SOLUTE CONCENTRATIONS IN g 100cm^{-3}		
		Protein	Glucose	Urea
A	1000	7.5	0.1	0.03
B	100	0.0	0.1	0.03
C	20	0.0	0.0	0.15
D	1	0.0	0.0	1.80

a) Explain how the following are brought about.
   i) The change in flow rate between B and D (2)
   ii) The change in protein concentration between A and B (2)
   iii) The change in glucose concentration between B and C (2)
   iv) The change in urea concentration between B and C (2)
   v) The change in urea concentration between C and D (2)
b) The smallest plasma protein molecules have molecular weights of about 69,000. Haemoglobin has a molecular weight of about 65,000. If red blood cells are damaged in the blood stream, haemoglobin may appear in the urine. Comment on the significance of these observations. (2)

(Total 12 marks)
(ULEAC)

Question 7.3
a) What is homeostasis? (2)
b) Describe the parts played by each of the following in the homeostatic mechanisms of the mammal:
   i) kidney (8)
   ii) liver (10)

(Total 20 marks)
(ULEAC)

Question 7.4
The diagram shows the blood supply to and from the liver.

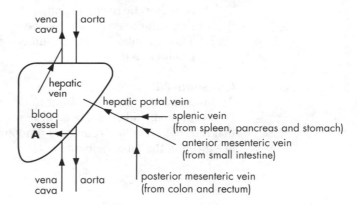

Fig 7.10

a) In which blood vessel would you expect to find the highest level of:
   i) glucose, 1 hour after eating
   ii) glucose, 10 hours after eating
   iii) carbon dioxide
   iv) insulin? (4)
b) i) What is the name of blood vessel A?
   ii) Name one substance that is required by the liver and is mainly supplied by this blood vessel. (2)

(Total 6 marks)
(AEB)

CHAPTER 7 EXAMINATION QUESTIONS

OUTLINE ANSWERS

Answer 7.1
i) E   ii) D   iii) D

Answer 7.2
a) i) Reduction in volume of fluid because of reabsorption. Widening of bore of tube.
   ii) Membrane of Bowman's capsule allows only proteins of molecular weight below 68,000 to pass. Plasma proteins have molecular weights of 69,000 upwards.
   iii) All glucose is reabsorbed in the proximal tubule by diffusion into the tubule cells.
   iv) Little urea is reabsorbed in the proximal tubule. 80% of the water is reabsorbed so the urea concentration increases five times.
   v) The volume of the filtrate falls to 1% of the original but about 50% of the urea is reabsorbed from the filtrate, accounting for the rise in urea concentration to only 1.80g 100cm^{-3}.
b) The haemoglobin molecules pass into the filtrate because they are smaller than the 68,000 threshold molecular weight. Normally they are in the red cells, which cannot pass through the membrane into the filtrate.

Answer 7.3
a) Maintenance of a **constant** internal environment; within narrow limits/independent of external conditions; dynamic process.
b) i) Urea/nitrogenous waste from body; by removal of urea from the blood; removal of uric acid; removal of creatinine; excretes glucose if blood level too high; regulates osmotic potential of blood; by control of water and mineral ions; role of ADH; role of aldosterone; regulation of blood pH; by excretion of hydrogen ions; by production of ammonium ions.
   ii) Removes excess glucose from blood; converted to glycogen under influence of insulin; converts glycogen to glucose if blood glucose level too low; under influence of glucagon; removes excess lipids from blood; transamination; deaminates **excess** amino acids; detoxifies poisons/drugs: e.g. hydrogen peroxide/alcohol; removes used haemoglobin/destroys old red blood cells; produces heat; eliminates used sex hormones; acts as blood reservoir.

Answer 7.4
a) i) anterior mesenteric vein   ii) hepatic vein
   iii) vena cava   iv) splenic vein
b) i) hepatic artery   ii) oxygen

QUESTIONS, STUDENT ANSWERS AND EXAMINER COMMENTS

Question 7.5
a) Explain the biological meaning of the terms:
   i) homeostasis   ii) negative feedback   *(2)*

> i) Homeostasis — is the maintenance of a constant internal environment, e.g. amount of glucose in blood. It involves self-adjusting mechanisms.
> ii) Negative feedback — is the process by which the correcting mechanism is triggered by the (thing) being regulated: e.g. increase in glucose begins the processes which decrease it.

❝ Correct — but not well expressed: 'change in level of metabolite' better than 'thing ❞

b) Alcohol may be considered as a drug. It has many effects on the body, including suppression of the anti-diuretic hormone (ADH) and promotion of vasodilation. Suggest an explanation for each of the following:
   i) a rise in blood alcohol level leads to an increase in urine formation
   ii) it is inadvisable to give alcohol to a person suffering from prolonged exposure to cold conditions.   *(4)*

CHAPTER 7 HOMEOSTASIS

> **Given in the question so superfluous**

> **More detail required for two marks. Essential points, ADH acts on kidney tubules, less water will be reabsorbed**

> **Should state arterioles**

> **Correct**

i) If the blood alcohol level rises, suppression of ADH occurs. ADH causes water to be reabsorbed in the kidney by making the cells lining the collecting ducts more permeable to water, so it can be drawn into surrounding blood vessels in the kidney.
   If less ADH is produced, due to alcohol intake, less water will be reabsorbed and larger quantities of more dilute urine will be produced.

ii) As alcohol promotes vasodilation, if it is given, the blood vessels in the skin may dilate, therefore letting more blood through. This will cause heat loss to the surroundings from the blood through the skin.
   If a person has been exposed to cold conditions, the blood temperature may fall dangerously if any heat is lost in this way.

(Total 6 marks)
(AEB)

Question 7.6

The diagram below shows some of the changes in the body that follow eating a meal:

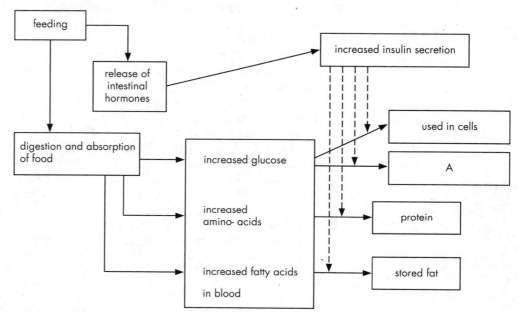

Fig. 7.11

a) Suggest an alternative fate for increased glucose which could be put in box A. *(2)*

> **Correct**

Increased glucose may be converted to glycogen and then stored in either the liver or the muscles.

b) What name is given to the groups of cells in the pancreas that secrete the hormone insulin? *(1)*

> **Correct**

The cells secreting insulin are the beta cells.

(1)

Indicate diagrammatically where the pancreas would fit into the scheme.

> Arrow from pancreas to increased insulin — correct

> Pancreas does not influence the release of intestinal hormones

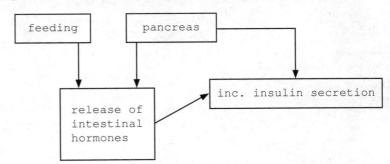

c) From the diagram give three factors which cause insulin secretion to decrease?
(3)

> Correct

```
Factors causing insulin secretion to stop include:
   A lowered level of glucose in the blood.
   A lowered level of amino acids in the blood.
   A lowered level of fatty acids in the blood.
```

(Total 7 marks)
(ODLE)

REVIEW SHEET

1 Briefly discuss the function of the kidney in terms of **excretion**.

_____
_____
_____
_____
_____

2 Write about each of the following processes involving the nephron function of the kidney.

a) Ultrafiltration: _____
_____

b) Selective reabsorption: _____
_____

c) Secretion: _____
_____

3 Complete the boxes in the following diagram with appropriate 'labels'. Then use the diagram to explain the operation of the **feedback system**.

_____
_____
_____
_____

4 Explain, with the aid of a diagram, the role of the ornithine cycle in homeostasis.

_____
_____
_____
_____
_____
_____

5 Contrast **ectothermic** animals with **endothermic**.

_____
_____

6 Explain the role of the **hypothalamus** in homeostasis for endothermic animals.

_____
_____
_____
_____
_____

7 Complete the following diagram and then use it to explain the control system for thyroxin secretion.

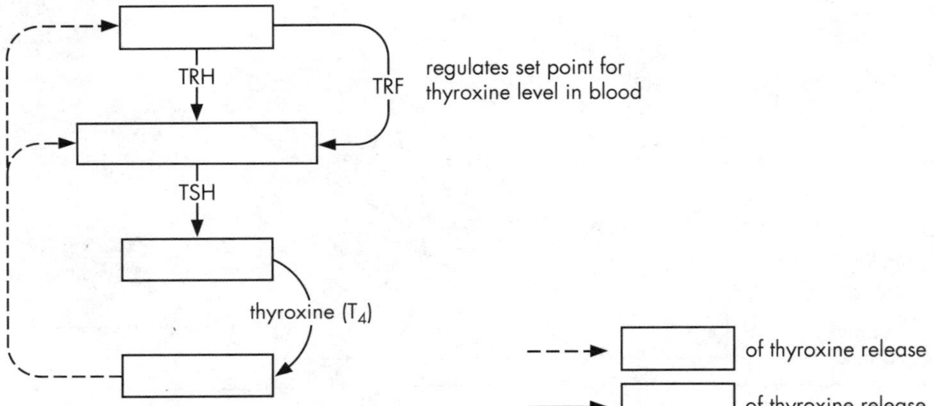

TRH   TRF   regulates set point for thyroxine level in blood

TSH

thyroxine (T_4)

- - - → [ ] of thyroxine release

⟶ [ ] of thyroxine release

_____
_____
_____

8 Briefly explain the mechanisms involved in the control of *water and salt balance* in the following lower animals.

a) Freshwater protozoa: _____

b) Insects: _____

c) Marine bony fish: _____

d) Freshwater bony fish: _____

CHAPTER 8

RESPIRATION AND GAS EXCHANGE

- THE ROLE OF ATP
- AEROBIC RESPIRATION
- ANAEROBIC RESPIRATION
- RESPIRATORY SUBSTRATES
- GAS EXCHANGE IN ANIMALS
- GAS EXCHANGE IN PLANTS
- PRACTICAL WORK

GETTING STARTED

Before beginning a study of respiration, it is sensible to appreciate the types and sources of the food materials which will be the respiratory substrates, so it is recommended that reference is made to the contents of Chapters 4, 5 and 6. In the sections on gas exchange, reference will be made to different organisms, so familiarity with Chapter 3 is also recommended.

This topic can involve some detailed biochemistry, but all syllabuses are very careful to specify exactly what is required of their candidates. If compounds are referred to by name, then these compounds should be learnt and their position in the metabolic process understood, but all syllabuses stress the amount of biochemical detail they require, and also that it is better to have a general understanding of the principles than a detailed knowledge of each reaction involved. You must check your syllabus for the exact wording of the section.

Respiration occurs in every living cell and is the only way in which cells can obtain **energy** for their activities. The process liberates **chemical energy** from food molecules; it is termed **aerobic** if it requires oxygen and **anaerobic** if it takes place in the absence of oxygen or if oxygen is not utilised.

There has been some confusion over the correct use of the term 'respiration' in the past, and you may find the reactions which take place inside the cells referred to as **internal**, **cell** or **tissue** respiration, to distinguish them from so-called **external** respiration, which involves the exchange of oxygen and carbon dioxide at the respiratory surface, more properly termed **gas exchange**.

CHAPTER 8 RESPIRATION AND GAS EXCHANGE

ESSENTIAL PRINCIPLES

THE ROLE OF ATP

The energy released when respiratory substrates are broken down is used to build up molecules of **adenosine triphosphate, ATP**, from **adenosine diphosphate, ADP**, and **inorganic phosphate**. When the molecules of ATP are **hydrolysed**, energy is released for reactions where it is needed in the cells (see Fig. 8.1).

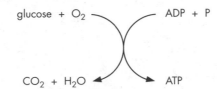

Fig. 8.1 ATP formation

ATP is found in all cells and is therefore the universal energy carrier. ATP molecules may be synthesised using energy derived from respiratory activity, a process known as **oxidative phosphorylation**, or from photosynthetic activity, **photophosphorylation**, in green plants. ATP has been shown to provide the necessary energy for:

- muscle contraction
- nerve transmission
- synthesis of materials within cells

AEROBIC RESPIRATION

Aerobic respiration can be summarised by the following equation:

$$C_6H_{12}O_6 + 6O_2 \rightarrow 6H_2O + 6CO_2 + 2880 \text{ kJ}$$

This summarises the process, indicating that a **carbohydrate substrate** is oxidised. **carbon dioxide** and **water** are produced as waste products and the total amount of energy yielded is **2800 kJ**. Other organic molecules are oxidised but we shall consider what happens to carbohydrates first, as they are normally used if available.

The breakdown occurs in two stages:

- the **glucose** is broken down to **pyruvic acid** in the process of **glycolysis**;
- the **pyruvic acid** is further broken down in the **mitochondrion** after it has crossed the mitochondrial membrane.

Glycolysis

The glucose molecule is first **phosphorylated** to make it more reactive using two molecules of ATP to form **hexose diphosphate**. This splits into two molecules of **triose phosphate (3C sugars)**, which are then converted to pyruvic acid, yielding four molecules of ATP and four hydrogen atoms; these are accepted by a **hydrogen acceptor molecule** called **nicotinamide adenine dinucleotide (NAD)**. At this stage there is a net gain of two ATP molecules. These reactions take place in the cytoplasm of the cell.

Pyruvic acid breakdown

Provided that oxygen is present, the pyruvic acid combines with **coenzyme A** to form **acetyl coenzyme A** and enters a mitochondrion; carbon dioxide is given off and the pyruvic acid loses two hydrogen atoms. The acetyl coenzyme A then enters the **tricarboxylic acid**, or **Krebs' cycle**, by combining with a 4C compound to form a 6C compound. This 6C compound undergoes reactions during which carbon dioxide and hydrogen atoms are removed until the 4C compound is regenerated and is ready to combine with another molecule of acetyl coenzyme A (see Fig 8.2).

Formation of ATP – the hydrogen carrier system

For each molecule of pyruvic acid that enters a mitochondrion, three molecules of carbon dioxide and ten hydrogen atoms are removed. The removal of the hydrogen atoms is carried out under the influence of **dehydrogenase** enzymes, and the

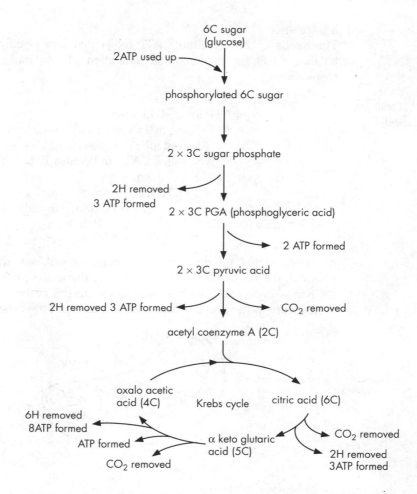

Fig. 8.2 Summary of aerobic respiration

hydrogen acceptor molecule is reduced. The hydrogen molecules are then passed on to a second acceptor molecule, which in its turn becomes reduced, leaving the first acceptor molecule oxidised. The process of oxidation and reduction is repeated with other carriers until the hydrogen atoms are eventually oxidised by molecular oxygen, forming water. The acceptor molecule that picks up the hydrogen atoms is NAD, and NAD hands on the hydrogen atoms to other carriers, each at a slightly lower energy level; sufficient energy is released to build up a molecule of ATP from ADP plus inorganic phosphate. For each pair of hydrogen atoms that passes along the chain of carriers, enough energy is released for the synthesis of three molecules of ATP if NADH involved; two if FADH. The sequence is as shown in Fig. 8.3.

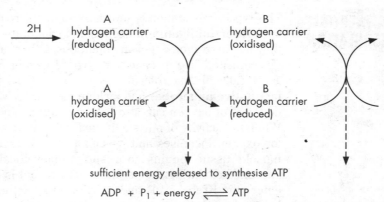

Fig. 8.3 Formation of ATP via the hydrogen carrier system

NAD and FAD

NAD and **FAD** are both **coenzymes** derived from the vitamin B complex, NAD from nicotinic acid and FAD from **vitamin B2**. The role of NAD and FAD in aerobic respiration shows us why these vitamins are so vital in the diet for the efficiency of the respiratory mechanism.

Balance sheet

The Krebs' cycle yields more ATP molecules than glycolysis. We can now draw up a balance sheet for the complete oxidation of one molecule of glucose in aerobic respiration:

> **66** Be familiar with this 'balance sheet' **99**

Glycolysis 2 ATP used
            4 ATP formed directly
            6 ATP (from two pairs of hydrogen atoms) picked up by NAD and
                passed along carrier chain
Net gain from glycolysis = 8 ATP molecules (if aerobic).
Krebs' cycle 2×1 ATP formed directly
                2×14 ATP from hydrogen carrier system (five pairs of hydrogen
                    atoms but one pair only yields 2 ATP)
Net gain from Krebs' cycle = 30 ATP molecules.
                Total = 38 ATP molecules.

As already stated, the reactions of glycolysis take place in the cytoplasm of the respiring cells, but the remaining reactions occur in the mitochondrion. Evidence suggests that the Krebs' cycle reactions take place in the **matrix** of the mitochondrion and the carrier systems are located on the **inner membrane** and the **cristae** (see Fig. 8.4)

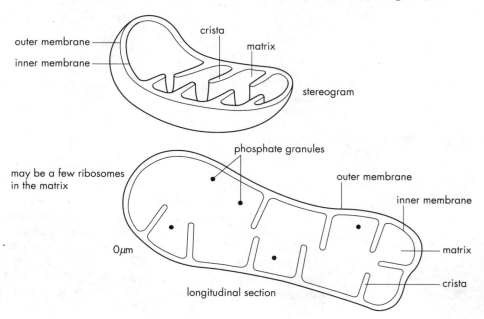

Fig. 8.4 Structure of a mitochondrion

ANAEROBIC RESPIRATION

This type of respiration is shown by some micro-organisms and by some tissues. Some bacteria are **obligate anaerobes** and cannot tolerate oxygen in their environment, so they live in oxygen-deficient conditions. Other micro-organisms, together with certain gut parasites, live in conditions where oxygen is not abundant, so they can respire aerobically or anaerobically as the conditions permit. These organisms are called **facultative anaerobes** and include the yeasts. Muscle cells which may be temporarily deprived of oxygen can also undergo anaerobic respiration for short periods of time. When the activity of muscle tissue is increased, as in a race or in exercise, the demand for oxygen increases and it cannot be delivered to the tissue quickly enough. The muscle tissue begins to respire anaerobically, producing **lactic acid**, which accumulates during the period of activity, but is either converted back to glycogen or enters the Krebs' cycle and is oxidised after the activity has ceased. The build-up of lactic acid constitutes an oxygen debt which has to be paid back later, when oxygen is again available (see Fig. 8.5).

$$CH_3 CO COOH + NADH + H^+ \longrightarrow CH_3 CH OH CO OH + NAD$$
    pyruvic acid   lactic acid

Fig. 8.5 Equation for anaerobic respiration in muscle

It is fairly obvious that without oxygen, oxidative phosphorylation will not take place so that the yield of ATP molecules will be greatly reduced. In the yeasts, the hydrogen atoms removed in glycolysis are taken up by acetaldehyde, which is converted to **ethanal** and carbon dioxide. The ethanal is converted to **ethanol**.

The following equation summarises the process:

> *A useful summary equation*

$$C_6H_{12}O_6 \rightarrow 2CH_3CH_2OH + 2CO_2 + 210 \text{ kJ}$$

This process is called **alcoholic fermentation** and is made use of in the brewing of beer and wine-making.

RESPIRATORY SUBSTRATES

CARBOHYDRATE

As has been mentioned earlier, carbohydrate in the form of **glucose** is the respiratory substrate in living cells. Carbohydrate is stored in plants in the form of starch, which can be converted to **sucrose** for transport in the phloem and to glucose for respiration. In animals, **glycogen** is the storage carbohydrate molecule. It is found in the liver and muscles; before it can be respired the glycogen must first be hydrolysed to glucose.

FAT

Fat is a good energy store and is used as a respiratory substrate when carbohydrate is in short supply. It has to be split into its constituent molecules of **glycerol** and **fatty acids**, first by hydrolysis and then the glycerol is converted into a 3C sugar which enters the Krebs' cycle yielding ATP molecules. The fatty acids go through a series of reactions in which 2C portions of the long hydrocarbon chains are split off as acetyl coenzyme A and enter the Krebs' cycle. This process takes place in the mitochondria and during the formation of the acetyl CoA, hydrogen atoms are removed, pass through the carrier system, resulting in the formation of ATP molecules. Very large numbers of ATP molecules are built up during this process, the precise number depending on the length of the hydrocarbon chain of the fatty acid.

PROTEIN

Protein is very rarely used as a respiratory substrate, usually only when all reserves of carbohydrate and fat have been used up. A certain amount of energy, though, is derived from the elimination of excess dietary protein. The protein is hydrolysed into its constituent amino acids and then it is **deaminated** in the liver. The amino group is converted into **urea** and excreted and the residue is converted to either acetyl CoA, pyruvic acid or some other Krebs' cycle intermediate, and oxidised.

Before leaving this section, many syllabuses expect candidates to carry out experimental work to show the action of dehydrogenase enzymes in yeast or germinating seeds. These experiments involve the use of indicators such as methylene blue, tetrazolium chloride (TTC) and DCPIP. You should check the relevant section of your syllabus for details.

RESPIRATORY QUOTIENT

It is possible to work out the **respiratory quotient** for a particular organism if we know the volume of carbon dioxide produced and the volume of oxygen taken up in a given period of time.

> *Know how to work out the respiratory quotient*

$$\text{Respiratory quotient (RQ)} = \frac{\text{volume of carbon dioxide produced}}{\text{volume of oxygen used}}$$

This information can give an indication of the type of food being used as the respiratory substrate by working out the theoretical RQ values for carbohydrate, fat and protein, and then comparing them with actual values obtained in experiments using

respirometers. Carbohydrates given an RQ of 1.0, fats about 0.7 and proteins 0.9. It must be borne in mind that most organisms do not use just one type of food as a respiratory substrate, so these exact values are seldom achieved. Man's RQ is usually about 0.85, indicating that a mixture of carbohydrate and fat is used for respiration.

GAS EXCHANGE

Gas exchange involves exchange of oxygen and carbon dioxide between the organism and its environment, and it takes place at a **respiratory surface** by the process of diffusion.

Respiratory surfaces must satisfy certain criteria if they are to carry out gas exchange effectively. The respiratory surface must be:

- **thin** so the diffusion paths are short
- **permeable** to the respiratory gases
- **moist** to allow gases to dissolve
- have a **large enough surface area** to satisfy the needs of the organism
- be close to the **circulatory system** in the majority of multicellular organisms so that the concentration gradients are maintained.

GAS EXCHANGE IN ANIMALS

PROTOZOA

In small organisms, such as the Protozoa, diffusion of gases occurs over the whole of the body surface. The surface area to volume ratio is such that it is large enough to satisfy the organism's needs, the diffusion paths are short and there is no need for a circulatory system.

EARTHWORMS

The earthworm is a much more complex but sluggish organism, and again the whole surface is used for the exchange of gases. The thin **epithelium** is kept moist by **mucus-secreting glands** and there is a good capillary network which removes the oxygen, thus maintaining steep diffusion gradients. The earthworm has the respiratory pigment **haemoglobin** dissolved in the plasma of the blood. This pigment increases the oxygen-carrying capacity of the blood.

INSECTS

With insects, the blood has no involvement in the carriage of oxygen, gas exchange occurring by means of a **tracheal system**. The second and third thoracic segments and the eight abdominal segments each bear a pair of openings called **spiracles** which lead into a series of tubes making up the tracheal system. The **tracheae** are lined with **chitin** to keep them open, and branch repeatedly to form **tracheoles**, which penetrate the tissues and organs of the insect's body. The tracheae and tracheoles ensure that oxygen is delivered directly to the respiring cells without the blood acting as intermediary. At rest, the ends of the tracheoles are filled with fluid in which the oxygen dissolves and diffuses to the tissues. When the insect is active, this fluid is drawn into the tissues because of an increase in lactic acid, which draws more air and thus more oxygen into the tracheoles when it is needed.

Large insects such as grasshoppers and locusts show ventilation movements of the abdomen. Air is taken in through the thoracic spiracles and pushed out through the abdominal spiracles by the abdominal segments' telescoping and extending (see Fig. 8.6).

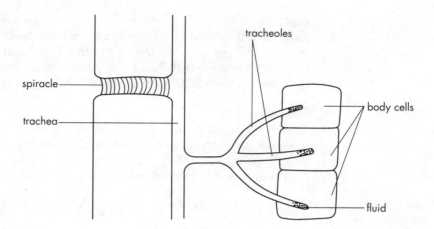

Fig. 8.6 Tracheal system in insects

BONY FISH

In a bony fish, gas exchange occurs over the **gills**, which are thin, permeable, have a large surface area and are well supplied with blood capillaries. Gills are made up of many **lamellae** and are covered by a protective bony plate, the operculum. When inspiration occurs, the **buccal cavity** expands due to the contraction of the muscles, which causes the floor of the buccal cavity to be lowered. The pressure within is decreased and water is drawn in through the mouth. At the same time the **opercular valve** is shut by the outside water pressure. The **opercular cavity** expands, drawing water through from the buccal cavity. This is helped by the contraction of the buccal cavity. Water is pushed over the gills and out through the operculum by inward movements of the operculum (see Fig. 8.7).

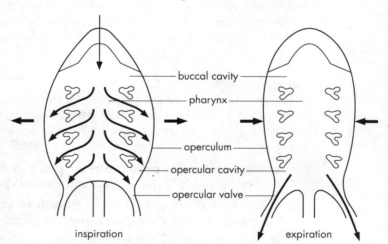

Fig. 8.7 Fish ventilation

MAMMALS

In a mammal, such as Man, the **thoracic cavity** contains the **lungs**, which are spongy in texture and consist of a branching network of tubes arising from a pair of **bronchi**. The tubes, or **bronchioles**, end in **atria**, which have large numbers of **alveoli**, providing a large surface area at which gas exchange takes place. The alveoli have a close association with the capillary network, so the gases do not have far to diffuse.

Inspiration
Inspiration is brought about when the **external intercostal muscles** contract, bringing the ribcage forwards and upwards. At the same time, the muscles surrounding the **diaphragm** contract, pulling the diaphragm flat. Both of these actions increase the volume of the thorax and the lungs but reduce the pressure inside the lungs so that it is below atmospheric pressure. Air rushes in down the trachea and into the lungs to equalise the pressure.

Expiration

Expiration then follows. The external intercostal muscles relax, restoring the ribcage to its former position, the muscles of the diaphragm relax, causing the diaphragm to resume its domed shape, and the volume of the thorax, and therefore the lungs, is decreased. This results in an increased pressure in the lungs and air is forced out.

Composition of air

The composition of **inhaled air** is the same as that of the **atmosphere**, but the **exhaled air** contains **more carbon dioxide** and **less oxygen**, together with more water vapour.

A comparison of inhaled and exhaled air is given in the table below:

> A useful comparison of inhaled and exhaled air

Table 8.1

	ATMOSPHERIC AIR (%)	EXHALED AIR (%)
Oxygen	20.95	16.4
Nitrogen	79.01	79.5
Carbon dioxide	0.04	4.1

Rate of respiration

At rest, a person breathing quietly will take in and expel about half a litre of air during a respiratory cycle, but as soon as any activity starts this **tidal volume** will be increased. The **rate of respiration** can be expressed as the volume of air breathed per minute, and it will vary with activity. Both the **frequency** and **depth** of inspiration will increase as activity increases. Not all the air that is taken in is used in gas exchange, as only that which comes into the alveoli that are in contact with the capillaries can be of any use. In the alveoli there is a thin film of moisture in which the oxygen can dissolve and then diffuse into the blood capillaries. The blood coming to the alveolus is about 70% saturated with oxygen, and when it leaves it is about 95% saturated. There is a concentration gradient between the oxygen and carbon dioxide in the blood and in the alveolus, so that as the blood leaves it has gained oxygen and lost carbon dioxide, but is in equilibrium with the gases in the alveolus.

Oxygen carriage

The **red blood cells** are important in the carriage of oxygen. They are filled with cytoplasm containing the red pigment **haemoglobin**. This pigment is a protein containing four **haem** groups, each with an iron atom and each attached to a polypeptide chain. Each haem group can combine with one molecule of oxygen.

- When the **partial pressure of oxygen is high**, as in the lung capillaries, oxygen combines with the haemoglobin to form **oxyhaemoglobin**.
- When the **partial pressure of oxygen is low**, as found in the respiring tissues then the **oxygen dissociates** from the haemoglobin.
- When the **partial pressure of carbon dioxide is high**, haemoglobin is less efficient at taking up oxygen and more efficient at releasing it. This is called the **Bohr effect**, after the man who discovered it (see Fig. 8.8).

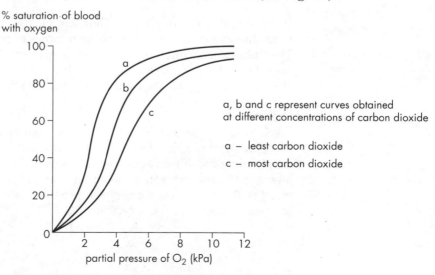

Fig. 8.8 Graph showing Bohr effect

Carbon dioxide carriage

The red blood cells are also involved in the way in which carbon dioxide is carried in the blood. Some of the carbon dioxide is carried in solution in the plasma, and some combines with the haemoglobin molecule to form **carbamino-haemoglobin**, but the majority of the carbon dioxide is transported as **hydrogencarbonate**.

The carbon dioxide diffuses into the red blood cells, where it combines with water to form **carbonic acid**. This dissociates into hydrogen carbonate and hydrogen ions under the influence of the enzyme **carbonic anhydrase**. The hydrogen ions cause the oxyhaemoglobin to dissociate into oxygen and haemoglobin. The oxygen diffuses out of the cell into the tissues, and the hydrogen ions associate with the haemoglobin, forming **haemoglobinic acid**. Hydrogen carbonate ions accumulate in the red blood cell, but steadily diffuse out because the membrane is permeable to them; to balance this outward movement of negatively charged ions, **chloride ions** diffuse in. This is known as the **chloride shift**; it maintains electroneutrality. The whole process is reversed when the red blood cells reach the lungs.

NERVOUS CONTROL OF BREATHING

Breathing is controlled by a **respiratory centre** in the **hindbrain**. This centre receives afferent nerves from the stretch receptors in the walls of the bronchioles, and these signal the degree of expansion of the lungs. As the lungs inflate, the frequency of the impulses increases and eventually inspiration is inhibited, so initiating expiration. The respiratory rate can also be influenced by an increased level of carbon dioxide in the blood, as happens during exercise, and by the higher centres of the brain. We can voluntarily alter our ventilation rate, thus overriding the involuntary mechanisms.

Before leaving this section, you are reminded that some syllabuses require a knowledge of the anatomy of the thorax of a mammal gained through dissection. In addition, experimental work using simple respirometers is suggested.

GAS EXCHANGE IN PLANTS

Gas exchange in plants occurs through the stomata and the lenticels. The **stomata** occur on the leaves and sometimes on the stems of herbaceous plants, while **lenticels** are found on the aerial parts of woody trees and shrubs.

Stomata

The stomata are pores in the **epidermis** of the leaf, bordered by two guard cells. The guard cells differ from the other epidermal cells in that they possess **chloroplasts**, and the inner wall of each one is thicker than the outer wall (see Fig. 8.9). They are distributed evenly on both sides of **isobilateral** leaves such as grass leaves, but usually occur in greater numbers on the underside of **dorsiventral** leaves such as apple leaves. Gases diffuse through the stomata along concentration gradients.

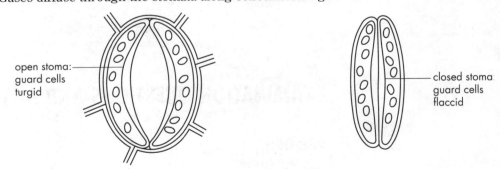

Fig. 8.9 Stomata

Lenticels

The lenticels are simply areas of loose cells in the bark, and diffusion of oxygen and carbon dioxide takes place through them relatively slowly. Oxygen diffuses in, and carbon dioxide plus some water vapour diffuses out. They are not of significance in the actively photosynthesising plant.

Diurnal cycle

CHAPTER 8 RESPIRATION AND GAS EXCHANGE

During the day, there is a **net uptake of carbon dioxide** and a **net output of oxygen** as photosynthesis is taking place rapidly, so any oxygen needed for respiration is obtained from the waste products of photosynthesis. At night, when there is no photosynthesis taking place, there is a net uptake of oxygen and a loss of carbon dioxide.

Stomatal mechanism

Many theories have been put forward to account for the opening and closing of the stomata. It is easy to observe that when the **guard cells** are turgid, they curve apart more and the pore between them widens; when they lose water, the pore becomes smaller. It is more difficult to explain why, for many plants, they open in daylight and close in darkness. The most recent theory to account for this suggests that **potassium ions** accumulate during the day, causing the uptake of water. On closure, the potassium ions diffuse out into the neighbouring epidermal cells, which results in water moving out of the guard cells.

PRACTICAL WORK

Suggestions for practical work have been made in the various sections of this chapter. It is worth emphasising that there is much practical work associated with the topics covered in the chapter. There are several experiments with dehydrogenase enzymes that are often set in practical examinations, and it is worth ensuring that both the respiratory activity and the enzyme involvement are thoroughly understood.

Experiments with simple respirometers which can measure oxygen uptake in a variety of small animals should be done. In some syllabuses, particularly where there are human physiology options, it is necessary to understand the workings of a simple spirometer.

Dissection as an examination exercise has been removed from nearly all A-level syllabuses. Some schemes of practical assessment do require students to perform dissections of a small mammal; the thoracic cavity is often mentioned as suitable. Whereas it is valuable to observe fresh material and to carry out a dissection, it is not always mandatory and it is likely that the requirement will disappear from most syllabuses as they are revised.

Other worthwhile practical work:

- epidermal strips of leaves – to investigate stomatal distribution
- observations on ventilation movements in locusts
- fermentation in yeast

Further reading

Calow P, 1981, *Invertebrate Biology*. Croom Helm
Marshall P T and Hughes G M, 1980, *Physiology of Mammals and Other Vertebrates*. CUP ABAL Unit 5, *Exchange and Transport*. Cambridge

EXAMINATION QUESTIONS AND ANSWERS

QUESTIONS

Question 8.1
a) What are the properties of a respiratory surface? *(4)*
b) Describe the mechanisms involved in the ventilation movements of the following animals:
  i) a terrestrial insect
  ii) a mammal *(10)*
c) Explain the roles of the following structures in gaseous exchange in plants:

i) stomata
ii) lenticels (6)
(Total 20 marks)
(ULEAC)

Question 8.2
The following passage is about cellular respiration. Read it carefully and then answer the questions.

The chemical required by living organisms in order to perform useful work is the result of a series of chemical reactions within the cell. The substance which acts as the energy source combines with phosphate to form a complex molecule which then undergoes a series of anaerobic changes known as glycolysis. If oxygen is present, a 2-carbon derivative of glycolysis is fed into a metabolic cycle in which oxidation reactions take place.

a) Name the substance which usually acts as the energy source in respiration. *(1)*
b) Name the substance which supplies the phosphate at the start of glycolysis. *(1)*
c) In what part of the cell does glycolysis take place? *(1)*
d) Name the end product of glycolysis. *(1)*
e) Name two processes which occur in living cells which require energy. *(2)*
(Total 6 marks)
(AEB)

Question 8.3
The diagram shows part of a gill filament from a fish.
a) What is the advantage of this flow arrangement? *(2)*

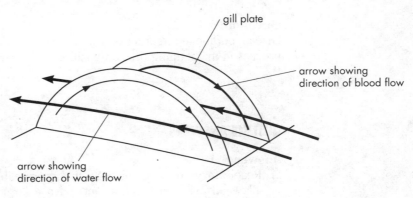

Fig. 8.10

b) Describe **two** ways, other than the flow arrangement, in which the gill filament is adapted as a respiratory surface. *(2)*
(Total 4 marks)
(AEB)

Question 8.4
The diagram on page 96 shows an outline of cellular respiration.
a) Name a polysaccharide commonly stored in:
   i) green plants
   ii) mammals *(2)*
b) i)   Name the process by which the 6-carbon sugar is converted to pyruvic acid.
   ii) Where in the cell does this take place?
   iii) Why is ATP used in this process? *(4)*
c) Name the compound formed from pyruvic acid in muscle cells under the conditions of oxygen debt. *(1)*
d) Name the type of enzyme involved at stage X. *(1)*
e) What happens finally to the hydrogen ions released from the tricarboxylic acid cycle? *(2)*
f) Make a labelled drawing of a mitochondrion, and on your drawing indicate where ATP synthesis occurs. *(4)*
(Total 14 marks)
(ULEAC)

CHAPTER 8 RESPIRATION AND GAS EXCHANGE

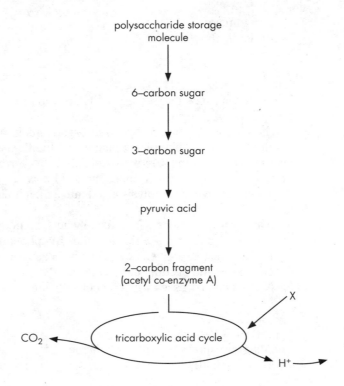

Fig. 8.11

Question 8.5
Give an outline of the process of aerobic respiration and discuss the significance of this process in maintaining the composition of the atmosphere.

(Total 20 marks)
(ULEAC)

OUTLINE ANSWERS

Answer 8.1
a) Respiratory surfaces must be thin, permeable to the respiratory gases, moist to allow diffusion in solution and have a large enough surface area to satisfy the respiratory needs of the organism. They need to be in direct contact with the respiratory medium (air or water) and close to a vascular system.

b) i) In an insect there are spiracles which open into a tracheal system. Large insects can move their abdominal segments by a pumping action. This is brought about by muscle action on the exoskeleton and enables the volume of the abdomen to be altered. When the volume is decreased, air is pushed out, and when it is increased air is taken in. Inhalation takes place through the front spiracles and exhalation through the abdomen. It is possible to see these movements in an insect such as a locust, together with the valves opening and closing the thoracic spiracles.

ii) In a mammal ribs articulate with the vertebrae and there are muscles between the ribs called intercostal muscles. When these muscles contract the ribcage is moved upwards and outwards in humans. At the same time the diaphragm muscles contract, causing the diaphragm to flatten. These two actions increase the volume of the thorax, so the pressure inside is reduced. Atmospheric air is at a higher pressure, so air is forced into the trachea, then into the bronchi and bronchioles. Exhalation involves the reverse of the process just described, aided by elastic recoil of the lung tissue. The whole process is controlled by the respiratory centre in the medulla of the brain.

c) i) Stomata are pores on leaves and green stems. They are surrounded by guard cells which cause the pore to be open when they are turgid. When the guard cells lose water and become flaccid, the pore closes. The pores are open during the day, allowing carbon dioxide into the leaf for photosynthesis.

ii) Lenticels are masses of loose corky cells found on woody stems. They allow gas exchange to occur. Oxygen can diffuse in and carbon dioxide diffuses out.

There may be some loss of water vapour. They are not involved in the uptake of carbon dioxide for photosynthesis.

Note. It is worth pointing out that b) of this question is about **ventilation** and not gas exchange at the respiratory surface, so students must resist the temptation to write all they know about the uptake of oxygen from the tracheoles of insects.

Answer 8.2
a) Glucose
b) ATP
c) Cytoplasm
d) Pyruvate (could put pyruvic acid or the chemical formula)
e) Chemical synthesis, nerve transmission, muscle contraction (any two)

Answer 8.3
a) The blood and water flow in opposite directions. This ensures that blood always meets water with a relatively higher oxygen content, maintaining a concentration gradient throughout the whole length of the gill filament.
b) The gill filament is sub-divided to increase the surface area for gas exchange; the tips of adjacent gill filaments overlap. This increases resistance to water flow, slowing the passage of water over the gill lamellae thus increasing the time available for gas exchange; there is a good blood supply.

Answer 8.4
a) i) Starch
   ii) Glycogen
b) i) Glycolysis
   ii) Cytoplasm
   iii) To phosphorylate the 6-carbon sugar and make it more reactive
c) Lactic acid
d) Dehydrogenases
e) Combine with electrons and oxygen to form water
f) Marks awarded for appropriately drawn double membranes, cristae and matrix, labelled correctly. An indication that ATP synthesis takes place on the cristae is needed.

Answer 8.5
This is any essay question and the first part of the answer requires much of the information which has already been given in answers to some of the questions above, together with more detail of what happens in the Krebs' cycle and the electron carrier system.

The reason this question has been included here is to stress that the second part needs to be answered in some detail, so careful thought and planning are necessary.

Reference should be made to the composition of the atmosphere and to the gas exchange which does go on between green plants and aerobic organisms to maintain the present equilibrium, before comments on how the balance can be upset by the activities of humans are made. It is relevant here to comment on the reduction of forests and plankton, and on the extra carbon dioxide being released by the burning of fossil fuels, but students must beware of losing their sense of proportion and haranguing the examiners about the evils of the greenhouse effect!

REVIEW SHEET

1. Explain, using a diagram, how ATP is formed.

2. What is the role of ATP?

3. Present an equation to summarise **aerobic respiration**.

4. Describe the process of **aerobic respiration** in words.

5. Complete the boxes with appropriate labels.

```
                    6C sugar
                    (glucose)
    ☐                  ↓
                       ↓
                       ☐
                       ↓
              2 × 3C sugar phosphate
                       ↓
    ☐ ←                ↓
              2 × 3C PGA (phosphoglyceric acid)
                       ↓
                       ↓                    ☐
              2 × 3C pyruvic acid
                       ↓
    ☐ ←                ↓                    → ☐
              acetyl coenzyme A (2C)
                       ↓
              oxalo acetic         citric acid (6C)
              acid (4C)   Krebs cycle
    ☐ ←
                         α keto glutaric
    ☐ ←                  acid (5C)          → ☐
              ☐
```

6 Present an equation to summarise **anaerobic respiration**.

7 Describe the process of **anaerobic respiration** in words.
_____
_____
_____
_____
_____

8 Present an equation to summarise **alcoholic fermentation**.

9 Explain what is meant by the **respiratory quotient** and outline its usefulness.
_____
_____

10 Outline the properties which **respiratory surfaces** must possess if they are to carry out gas exchange effectively.
_____
_____
_____

11 Outline the process of gas exchange, using an appropriate diagram to illustrate your answer.
  a) Gas exchange in insects:
  _____
  _____
  _____
  _____
  _____
  _____
  _____

  b) Gas exchange in bony fish
  _____
  _____
  _____
  _____
  _____
  _____
  _____

12 Outline these processes in mammals.
  a) Oxygen carriage: _____
  _____
  _____
  b) Carbon dioxide carriage: _____
  _____
  _____

CHAPTER 9
CHEMICAL CO-ORDINATION IN PLANTS

PLANT GROWTH SUBSTANCES

PLANT MOVEMENTS

EFFECT OF LIGHT ON PLANT GROWTH

PHOTOPERIODISM

PRACTICAL WORK

GETTING STARTED

This section of the syllabus concentrates on the effect of **hormones** on the growth of plants, covering the breaking of dormancy in the seed, the direction in which plant organs grow and the control of flowering.

Germination, the activities of apical and lateral meristems and the processes of cell elongation and differentiation are included in Chapter 13. The emphasis in this chapter is on the factors which influence and control growth.

In animals chemical co-ordination is achieved by means of hormones, which are produced in small amounts in one part of the body and have their effect in a different part of the body. Because co-ordination in plants is achieved by chemicals which do not always move far from the site of synthesis, they are often referred to as growth substances rather than hormones. It is not always clear exactly how these **growth substances** bring about their effects, but they are essential to the processes of **cell division**, **cell elongation** and **cell differentiation**.

There are five major types of plant growth substances:

- **auxins** and **gibberellins**, which are mainly concerned with cell elongation and differentiation
- **cytokinins**, which are concerned with cell division
- **abscisic acid**, which is concerned with dormancy
- **ethene**, which is concerned with senescence

CHAPTER 9 CHEMICAL CO-ORDINATION IN PLANTS

ESSENTIAL PRINCIPLES

PLANT GROWTH SUBSTANCES

AUXINS

Early experiments on auxins

Early experiments on auxins used **oat coleoptiles** and investigated the effects of light on the direction of growth.

- If oat coleoptiles were subjected to unilateral illumination, they showed a positive **phototropic response** and grew towards the direction of the light, the bending occurring just behind the tip.
- If the tip of the coleoptile was removed, then no response was shown to the direction of the light, and similarly no response was shown if the tips were covered.
- If the coleoptile was covered except for the tip, then a positive phototropic response was again shown.

As a result of these experiments, it could be concluded that the tip detected the stimulus of light and that the response was made in the region just behind the tip (see Fig. 9.1).

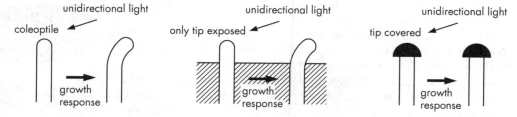

Fig. 9.1 Effects of light on growth of oat coleoptiles

Further experiments

Further experiments were done to find out how the stimulus was transmitted to the region which responds. A thin sheet of impermeable material, such as mica, was inserted into the coleoptile just below the tip; the coleoptiles were exposed to unilateral illumination.

- If the mica is inserted on the illuminated side, then curvature occurs.
- If it is inserted on the unilluminated side then no curvature occurs.

This suggests that there is some growth-promoting substance produced at the tip which is prevented from moving down the dark side of the coleoptile by the mica insert (see Fig. 9.2).

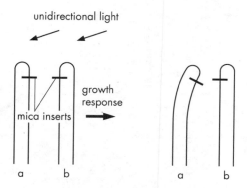

Fig. 9.2 Effects of light on oat coleoptiles with mica inserts in tips

Isolation of auxin

Attempts were then made to collect this substance by removing the tips of coleoptiles and placing them on small blocks of agar. The tips were left for several hours so that the chemical substance could diffuse into the agar blocks, then the blocks were placed on decapitated coleoptiles. The effects were compared with similar experiments using untreated agar blocks and the results are shown in Fig. 9.3.

CHAPTER 9 ESSENTIAL PRINCIPLES

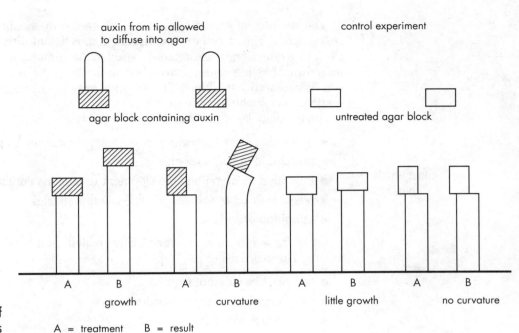

Fig. 9.3 Collection and effects of auxin on decapitated coleoptiles

How auxin exerts its effect

The chemical substance was called 'auxin', identified as **indoleacetic acid (IAA)**, and subsequently was found to occur in many plants. It is now known that IAA exerts its effect by enabling the cellulose microfibrils to slide past each other so that the cell wall can be stretched more easily as the cell takes up water osmotically during elongation. Auxin is produced by the tips of the coleoptiles and stem tips and diffuses downwards. When coleoptiles are exposed to unilateral light, the auxin diffuses away from the illuminated side to the darkened side, where it causes greater elongation of the cells, thus bringing about curvature towards the source of light (see Fig. 9.4).

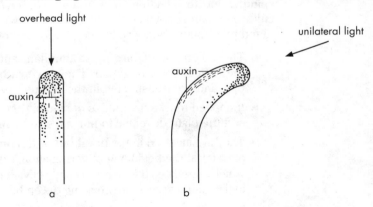

Fig. 9.4 Distribution of auxin in coleoptile tips

Geotropic response

Roots respond to **gravity** by growing downwards, which is a positive **geotropic response**, while **shoots** respond by growing upwards and are negatively geotropic. This can be demonstrated quite simply using a **klinostat**, consisting of a rotating drum to which seedlings can be fixed. When this drum rotates it eliminates the effects of gravity by exposing all sides of the seedlings equally to the stimulus, so the root and shoot will grow straight, but when it is stopped, the root and shoot will show their normal responses. It has been shown that high concentrations of auxin will inhibit cell elongation in roots, so in a root placed horizontally, where auxin accumulates on the lower surface, elongation will occur on the upper side, causing the root tip to curve downwards.

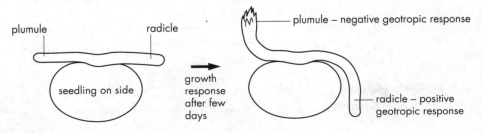

Fig. 9.5 Growth response of seedling on its side

The stimulus of gravity appears to be detected by the movement of starch grains in certain cells. The starch grains are called **starch statoliths** and they come to rest on the lower sides of cells in organs exposed to the stimulus of gravity. In some way, as yet unknown, they affect the distribution of auxin in the region of the root apex. More recent research has shown that another growth substance may be involved in the response shown by roots to gravity.

Auxins do have other effects on plant growth:

> **Other effects of auxin**

- promotion of root growth from cuttings (formation of adventitious roots on stems)
- promotion of fruit growth
- inhibition of lateral bud development (maintains dominance of the apical bud)
- possible delay of leaf senescence in a few species
- inhibition of leaf fall

Many of these effects are commercially applied, as it is possible to produce synthetic auxins. These are used:

- to help the setting of fruit
- as rooting hormones in powder form
- as selective weedkillers

2,4-dichlorophenoxyacetic acid (2,4-D) is a synthetic auxin which is used to kill broadleaved plants in cereals crops and on lawns. '**Agent Orange**' is a mixture of 2,4-D and another synthetic auxin, 2,4,5-T, and was the defoliant used in Vietnam by the US forces.

GIBBERELLINS

This group of plant growth substances was discovered in Japan, when investigations on a fungus disease of rice were being carried out. The rice plants grew very tall and spindly, due to infection by a fungus, *Gibberella*. The substance extracted from the culture medium on which the fungus had grown was called **gibberellin**, and it was later found that many plants produced gibberellins of their own.

- These compounds are most abundant and active in young plant organs, where they affect cell elongation. They are synthesised in young leaves, buds and root tips, and are transported in the vascular tissues.
- If applied to dwarf varieties of plants, they stimulate the rapid growth of the stem and the plant develops to normal size by increasing internode growth.
- They are involved in the **breaking of dormancy** in buds and seeds: after the uptake of water at the beginning of germination, the embryo begins to produce gibberellin, which stimulates the synthesis and activation of enzymes involved in the breakdown of the food reserves, so providing the embryo with materials for growth (see Fig. 9.6).

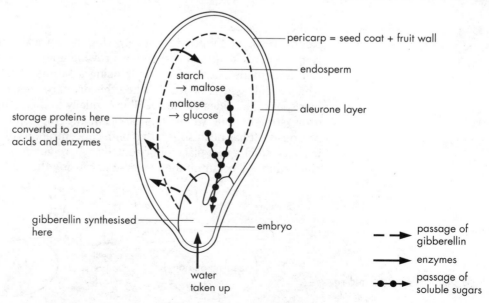

Fig. 9.6 Gibberellin involvement in germination

- Gibberellins have no effect on root growth, nor are they involved with tropisms, but they do interact with other growth substances. They can be produced commercially from fungi and are used to promote the setting of fruit and in parthenocarpy.

CYTOKININS

The cytokinins form a group of plant growth substances, first extracted from coconut milk, which are involved with **cell division**. They occur in very small quantities, especially where cell division is occurring. Together with auxin, they promote cell division in the apical and lateral meristems, providing an excellent example of the interaction of growth substances in plants. They have also been shown to cause delay in the ageing of leaves, and are involved in the breaking of dormancy in buds and seeds.

ABSCISIC ACID

Abscisic acid is a **growth inhibitor** and appears to be antagonistic to the action of the other growth substances. It is synthesised in stems, leaves, fruits and seeds and is transported in the phloem. It promotes dormancy in buds and seeds, and is also involved in fruit and leaf fall.

ETHENE

Ethene is a gas which is made by nearly all plant organs and is known for its effect on the ripening of fruit. It appears to stimulate the rapid increase in respiration rate which occurs before the ripening of fruit. Knowledge of its effects has been used in the commercial control of ripening. Fruit is stored in oxygen-free conditions to prevent ripening, which can be induced by applying oxygen and ethene when required.

Interaction of growth substances

There is a great deal of evidence that growth substances do not work on their own but work by interacting with one another, either by supplementing each other or by having an antagonistic effect. Gibberellins and auxins supplement each other in stem elongation, and we have seen that cytokinins need auxins in order to promote cell division. There is also evidence that gibberellins are involved in the division of the cambium. For further up-to-date information on this topic, you are recommended to refer to articles in scientific journals and periodicals.

PLANT MOVEMENTS

Entire plants do not usually show movement, but individual organs can respond to external stimuli. Movements can be divided into three types:

- tropic
- nastic
- tactic

TROPIC MOVEMENTS

A **tropic movement** is made by a plant organ in response to a **unidirectional stimulus**; the direction of the movement is related to the direction of the stimulus. The movement is achieved by the growth of the plant organ concerned.

Phototropic and geotropic movements

Phototropic movements and geotropic movements have already been referred to in this chapter, and their mechanism described, but some syllabuses do require a knowledge of other tropic movements.

Chemotropic movements
A **chemical stimulus** causes chemotropic movements; this can be shown by pollen tubes growing towards a chemical produced by the micropyle of the ovule.

Hydrotropic movements
Some roots can be shown to be positively hydrotropic by growing towards **water**.

Aerotropic movements
Pollen tubes show negative aerotropism, growing away from **oxygen**.

Thigmotropic movements
Thigmotropic movements are shown by the tendrils of peas as they twine round a solid support.

NASTIC MOVEMENTS

A **nastic movement** is a **non-directional** movement of a plant organ in response to a diffuse stimulus. The movement may be due to growth or to changes in turgor. Flowers and leaves respond to changes in light intensity and temperature by opening and closing, or showing 'sleep movements'. For example, tulip petals close at night and open during the day, in response to more rapid growth on the lower and upper surfaces of the petals respectively, triggered off by temperature changes. Another example is the rapid drooping of the leaves of *Mimosa pudica* when touched. This is not a growth movement; it is due to **changes in turgor**, thought to be transmitted by a hormone.

TACTIC MOVEMENTS

Tactic movements are shown by whole cells or organisms and are characteristic of gametes and unicellular plants. The response is a directional one made in response to a **directional stimulus**, either positively towards the stimulus or negatively away from it.

EFFECT OF LIGHT ON PLANT GROWTH

Light affects many aspects of plant growth, from the synthesis of chlorophyll to photosynthesis and phototropic movements. A plant that grows in total darkness does not develop chlorophyll and grows spindly with long internodes and poorly developed leaves. It is said to be **etiolated** and will die as soon as all the food reserves are used up.

PHOTORECEPTOR – PHYTOCHROME

As a result of a great many experiments, including work on the effect of light on the germination of certain types of lettuce seeds, it was suggested that some kind of pigment existed in plants which was responsible for absorbing the light, a **photoreceptor**. This photoreceptor has been identified as **phytochrome**, a blue-green pigment found in very small quantities in plants. It is suggested that phytochrome is activated by appropriate wavelengths of light and that it then causes a hormone to be produced which influences cell division or elongation.

Phytochrome exists in two forms which are interconvertible.

- One form absorbs red light, with an absorption peak at 665 nm, and is referred to as P_R or P_{665}.
- The other form absorbs far red light, with an absorption peak at 725 nm, and is referred to as P_{FR} or P_{725}.

When P_R absorbs red light it is converted to P_{FR} and vice versa. There is more red light than far red light in natural sunlight, so P_{FR} tends to accumulate during the day and becomes slowly converted back to P_R at night.

It has been shown that the phytochrome system is important in many aspects of plant growth, such as germination in some seeds, stem elongation, leaf expansion, growth of lateral roots and flowering.

PHOTOPERIODISM

Photoperiodism is the term used to describe the influence of the relative lengths of periods of light and darkness on the activities of plants, and day-length has been shown to have an important effect on flowering. Flowering plants can be divided into three groups according to their photoperiodic requirements prior to the production of flowers.

> Useful groups for flowering plants

- **Day neutral plants** Flowering does not seem to be affected by the day length, e.g. tomato, cotton, cucumber.
- **Long day plants** Flowering is induced by exposure to dark periods shorter than a critical length, e.g. cabbage, antirrhinum, petunias.
- **Short day plants** Flowering is induced by exposure to dark periods longer than a critical length, e.g. chrysanthemums, tobacco, poinsettias.

Flowering in short day plants is inhibited by exposure to red light, and exposure to far red light will bring about flowering. It seems that these plants will flower only if the level of P_{FR} is low enough, but the situation in the long day plants is reversed and flowering is triggered of by high levels of P_{FR}.

The photoperiodic stimulus is perceived by the leaves and this can be shown easily by leaving one leaf of a plant exposed while the rest of the plant is covered up. The stimulus must be transmitted in some way to the buds, which then develop flowers. The suggestion is that the phytochrome causes a hormone to be produced which affects the buds. This hormone has been named '**florigen**' but has not so far been isolated, and it is more likely that some of the growth substances already discussed in this chapter are involved.

PRACTICAL WORK

There is not a great amount of practical work that can be carried out easily by the A-level student on this section of the syllabus. Experiments can be devised using auxin in lanolin paste, and it is not difficult to show the effects of unidirectional light and gravity on shoots and roots. A klinostat is a fairly simple piece of apparatus and can be used to eliminate the effects of gravity and unidirectional light in simple experiments with potted plants or seedlings. It is also possible to demonstrate the photoperiodic response by changing the relative lengths of day and night. The experiments mentioned here are usually set up as demonstrations as it can take several weeks to achieve a result.

It must be remembered that the implications of the effects of the growth substances on plants are important in agriculture and horticulture.

Further reading
Street H E and Opik H, 1984, *The Physiology of Flowering Plants*. Arnold.

EXAMINATION QUESTIONS AND ANSWERS

QUESTIONS

Question 9.1
a) What is *photoperiodism*? *(3)*
b) How does the photoperiod affect:
   i) flowering *(7)*
   ii) dormancy in plants *(4)*
   iii) breeding behaviour in animals? *(6)*
   (Total 20 marks)

Question 9.2
An investigation was carried out into the effect of gibberellin on the growth of leaves in dwarf bean plants. Equal amounts of the hormones were applied either to the stem or

to the first leaves produced by the plants. In one experiment the plants were left intact, but in a second experiment the growing point (apex) of each plant was removed when gibberellin was applied. In both experiments a control group of plants received no gibberellin.

By reference to the graphs, answer the following questions:

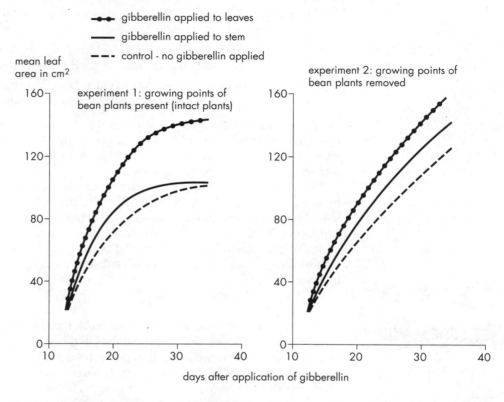

a) Compare the effects of applying gibberellin to the stem and to the leaves of the intact bean plants. *(4)*

b) i) Describe the effect on leaf growth of removal of the growing points from the bean plants when no gibberellin is applied. *(2)*

   ii) Suggest **one** reason why removal of the growing point has this effect. *(1)*

c) i) How did removal of the growing points from the bean plants affect the results of gibberellin application? *(3)*

   ii) Suggest **one** hypothesis which could explain this effect. *(2)*

   iii) Outline **one** additional experiment to test this hypothesis. *(3)*

d) i) Suggest **two** cellular mechanisms which may lead to an increase in leaf area. *(2)*

   ii) Suggest a mechanism by which a hormone such as gibberellin may exert its effect on leaf growth. *(1)*

e) Give **one** similarity and **one** difference between a plant hormone such as gibberellin and a typical animal hormone. *(2)*

(Total 20 marks)
(ULEAC)

OUTLINE ANSWERS

Answer 9.1

a) Definition required here; see chapter.

b) i) Reference to long day, short day and day neutral plants together with an explanation of the link with phytochrome and how flowering is brought about in both long day and short day plants is required. Give named examples where possible.

   ii) This topic has not been covered in detail in this chapter, but it is believed to be brought on by shorter days in autumn. This is thought to be linked to a shortage of either auxin or gibberellin, and only when the hormone levels are built up again is dormancy broken. Dormancy may be brought on by accumulation of a growth-inhibiting hormone, such as abscisic acid, which must then be removed or decrease before growth can be resumed. Experiments with auxin and gibberellin have shown that dormancy can be delayed if plants are treated with these hormones.

iii) This section is not relevant to this chapter but to the topic of photoperiodism generally. The photoperiod is the major external factor regulating rhythmic activity in animals, and breeding behaviour is influenced by day length. Activities such as nest-building in birds and copulation in mammals can be linked to the photoperiod, whether it be shorter days or longer days, so that the young are born at the most favourable time of year.

Answer 9.2
a) Applying gibberellin to the leaves and to the stem gives a higher growth rate; greater overall expansion of leaves when applied to leaves; with stem application leaf area levels off; does not increase area that much; with leaf application, leaves still expanding after 35 days.
b) i) Removal of growing points causes greater leaf expansion; no plateau reached after 35 days.
   ii) Hormone or auxin made at the tip suppresses leaf expansion.
c) i) The expansion rate was lower at first, but went on for longer; it did not plateau. The final leaf area was greater.
   ii) The hormone at the tip inhibits the effect of gibberellin, or it may affect the transport of the gibberellin to the leaves.
   iii) Gibberellin and auxin could be applied to one decapitated shoot, and only gibberellin to another one. The one with gibberellin only would show the greatest expansion.
d) i) Mitosis; cell expansion.
   ii) Stimulating cell expansion.
e) The hormones are similar in that they are both organic chemicals effective in small quantities. They are different in that animal hormones are made in special endocrine glands.

QUESTIONS, STUDENT ANSWERS AND EXAMINER COMMENTS

Question 9.3
Flowering is not a random process; both wild and cultivated plants grown outdoors produce flowers at particular times of year.

In an experiment, the upper leaves of a chrysanthemum plant (A) were removed but the growing buds and lower leaves were left intact. The upper half of the plant was then exposed to long days and the lower leaves to short days. The plant flowered.

The experiment was repeated using a similar plant (B), but exposing the lower leaves to long days and the defoliated upper half to short days. This plant did not flower.

The results of the experiments are summarised in the diagrams below.

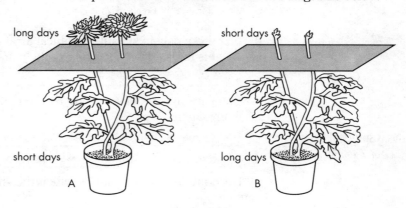

CHAPTER 9 CHEMICAL CO-ORDINATION IN PLANTS

a) i) What conclusions can you draw from these experiments?
   The light stimulus is detected by the leaves, and a substance is made in the leaves which makes the plant flower. (4)

> **The first sentence is correct. The substance should be referred to as a hormone perhaps but student would not be penalised for this. But more is needed; some reference to the hormone being transmitted to the apex where it controls flowering. This answer is too brief for 4 marks**

ii) What name is given to the effect described? **Correct answer**
   photoperiodism (1)

iii) Further experiments have shown that if the dark period is interrupted, by even a short period of light, flowering is prevented. What does this suggest?
   The substance works best in the dark. (2)

> **Not really the point; the hormone is made in the dark; destroyed by light. It really means that the dark period length is more important than the light period length**

b) Chrysanthemums are garden perennials which are much valued for their flowers, but unfortunately, in the garden, they have a limited flowering period. Suggest how the production of chrysanthemum flowers can be undertaken commercially throughout the year.
   You could keep plants in greenhouses where you could give them the right conditions of light and dark. (3)

> **This is fine as far as it goes, but the temperature and other conditions can also be controlled and these are important in producing good flowers all the year round. This answer too brief as well**

c) Explain why chrysanthemum growers commonly pinch out the terminal buds of their plants several times in each season.
   To get bigger flowers. (2)

(Total 12 marks)
(ULEAC AS-level)

> **The student does not have much idea here and has missed the point. Apical bud inhibits production of lateral buds; if lateral buds grow the plants become bushy and lots of flowers are produced.**

Question 9.4

A pea seedling with a straight plumule and radicle was illuminated from one side for a period of 8 hours. Measurements were made of the angle of curvature of the radicle and the plumule at 2-hourly intervals during the 8-hour period.

The results are shown in the table below.

LENGTH OF EXPOSURE TO UNILATERAL ILLUMINATION /HOURS	ANGLE OF CURVATURE	
	Plumule	Radicle
0	0°	0°
2	+7°	−3°
4	+14°	−4°
6	+18°	−8°
8	+23°	−11°

Source: Adapted from P W Freeland, *Problems in Theoretical Biology*

a) i) Plot these data in suitable graphical form on the graph paper opposite. (5)
   ii) What is the angle of curvature of the plumule after 3 hours' illumination?
   10.6° (1)

> **Correct**

b) i) Compare the responses made by the plumule and the radicle to the unilateral illumination.
   The plumule is positively phototropic and curves towards the light. The radicle is negatively phototropic and curves away from the light. (2)

> **This is sufficient detail for 2 marks**

ii) Describe the mechanism by which the stimulus results in the response of the plumule.

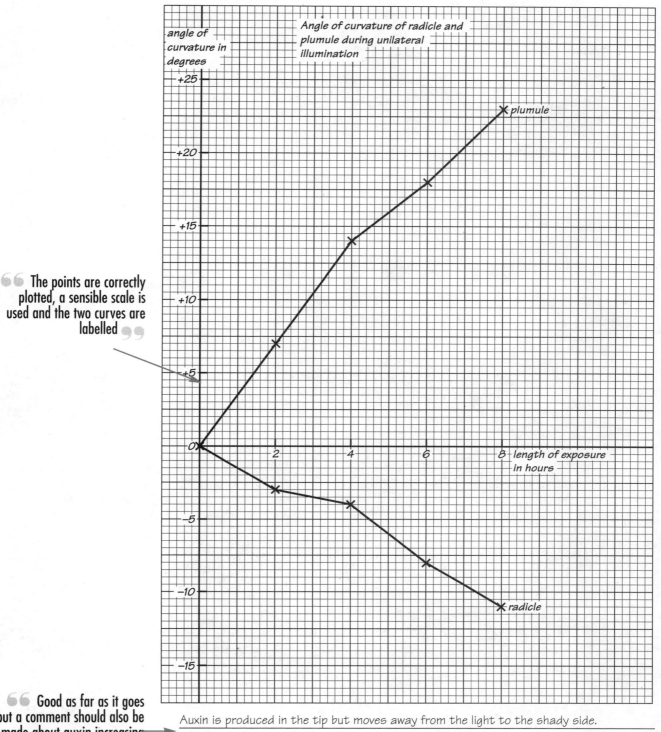

> **The points are correctly plotted, a sensible scale is used and the two curves are labelled**

> **Good as far as it goes but a comment should also be made about auxin increasing cell elongation or affecting the cell wall**

Auxin is produced in the tip but moves away from the light to the shady side. This causes more growth on the shady side and so the shoot curves towards the light.
(4)

c) Suggest how the responses of the plumule and the radicle are important in the successful growth of the seedling.

> **A good qualified comment**

Curvature of the plumule towards the light puts the leaves in a good position for photosynthesis. Curvature of the radicle from the light means that it would grow into the soil.
(2)

> **This should go on to comment on why the radicle needs to be in the soil, that is, a comment about anchorage or absorption of nutrients**

(Total 14 marks)

NB Review Questions on this chapter can be found at the end of Chapter 10.

CHAPTER 10
CO-ORDINATION IN ANIMALS

GETTING STARTED

Unlike plants, co-ordination in animals is brought about by means of two separate but interconnected systems, the **nervous system** and the **endocrine system**. The former gives a rapid and short-term response whilst the latter gives a slower but more lasting one. This chapter, in line with the examination syllabuses, deals mainly with co-ordination in mammals. Animal behaviour is related to co-ordination in mammals. Animal behaviour is related to co-ordination as it is based on the interpretation of response to sensory inputs. Cross-reference should be made to homeostasis, Chapter 7, and also to the chapters dealing with other topics in mammalian physiology, especially transport, locomotion, respiration and nutrition.

ORGANISATION OF NERVOUS SYSTEMS

VERTEBRATE NERVOUS SYSTEMS

STRUCTURE OF NEURONES

TRANSMISSION OF NERVE IMPULSES

SYNAPSES

SENSORY RECEPTORS

ENDOCRINE GLANDS AND THEIR SECRETIONS

ANIMAL BEHAVIOUR

PRACTICAL WORK

CHAPTER 10 CO-ORDINATION IN ANIMALS

ESSENTIAL PRINCIPLES

ORGANISATION OF NERVOUS SYSTEMS

MULTICELLULAR ORGANISMS

All multicellular organisms have nervous systems formed of specialised nerve cells, **neurones**, which transmit impulses.

Hydra
The simplest type is found in cnidarians such as *Hydra*. It consists of a double nerve net, one each side of the **mesogloea**. Most of the neurones synapse with others but some may be in direct connection. *Hydra* is sensitive to touch, noxious chemicals and food. Impulses pass through the nerve net and movement is produced away from the former two and towards the last (see Fig. 10.1).

Earthworms
More complex systems are exemplified by annelids such as earthworms. These have a central nervous system which consists of a double ventral nerve cord terminating at the anterior end with a **sub-pharyngeal ganglion**. Above the pharynx is a pair of **cerebral ganglia** joined to the nerve cord by a pair of **circumpharyngeal commissures** (see Fig. 10.2). Nerves pass out from the ganglia and nerve cord to supply the various segmental organs and to bring stimuli into the central nervous system. Earthworms have no major sense organs but have sensory cells scattered in the epidermis and muscles. Most other invertebrates have a similar pattern of nervous system. The exception is the molluscs, whose nervous system and sense organs are more like those of vertebrates.

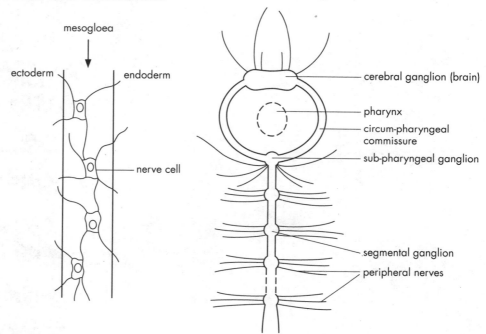

Fig. 10.1 Hydra nerve net   Fig. 10.2 Earthworm nervous system

VERTEBRATE NERVOUS SYSTEMS

These reach their highest development in mammals, which will be used as the type example. A vertebrate nervous system for descriptive purposes is divided up into a central nervous system of **brain** and **spinal cord** and a peripheral system of **sensory** and **motor** nerves. Anatomically associated with the motor nerves but functionally separate is the **autonomic system**.

STRUCTURE AND FUNCTION OF THE BRAIN

The brain is divided into three main regions, **forebrain**, **midbrain** and **hindbrain** (see Fig. 10.3).

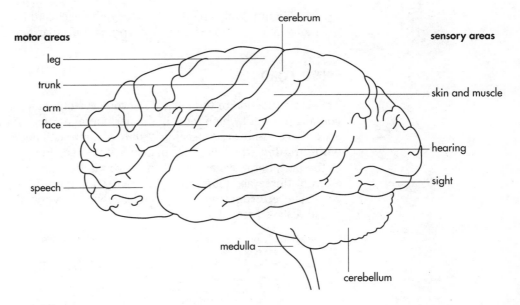

Fig. 10.3 Surface view of a human brain

Forebrain

The dorsal part of the forebrain consists of two large cerebral hemispheres forming the **cerebrum**. This consists of an outer, highly folded layer made up of nerve cells and an inner region of nerve fibres. All the higher mental activities take place in the cerebrum. These include the interpretation of sensory information, thought processes and memory. There are specific areas behind the central **sulcus** to receive inputs from different parts of the body and equivalent motor areas in front of it sending instructions to these same regions. Other regions of the cerebrum deal with personality, speech, hearing and sight. In the centre of the forebrain is the **thalamus**, which relays information to and from the cerebrum. Beneath the thalamus is the **hypothalamus**. This contains centres which control the basic drives and the emotions. Water balance and temperature regulation are also controlled by this structure and it produces hormones which are stored in the anterior pituitary (see Fig. 10.4).

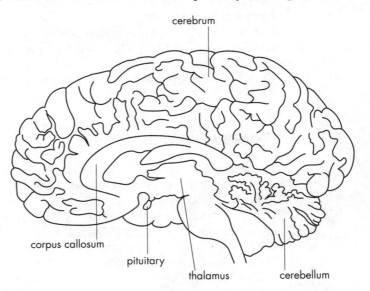

Fig. 10.4 Sagittal section of a human brain

Midbrain

The midbrain is relatively small in mammals and acts to link the forebrain and the hindbrain.

Hindbrain

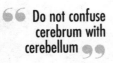

Do not confuse cerebrum with cerebellum

The hindbrain consists of two main parts, the **medulla** and the dorsal **cerebellum**. The medulla contains the **cardiac, respiratory** and **vasomotor** centres, which control vital body activities; it is the origin of much of the autonomic nervous system. The cerebellum is concerned with the co-ordination of body movements and postural responses. Impulses from the motor centres in the forebrain pass to the spinal cord via the cerebellum. It also receives sensory impulses and has the function of moderating the motor impulses to produce appropriate muscular movements.

SPINAL CORD

This is the portion of the central nervous system which passes through the vertebral column and from which most of the peripheral nerves originate. It consists of a central area of **grey matter**, consisting mainly of nerve cell bodies, and a surrounding area of **white matter**, which consists of nerve fibres (see Fig. 10.5). **Afferent** (sensory) fibres of the peripheral system enter the cord via the **dorsal roots**. The cell bodies of the afferent fibres are found in the **dorsal root ganglia**, which lie alongside the spinal cord. The **efferent** (motor) fibres leave via the **ventral roots**. There are 31 pairs of spinal nerves present.

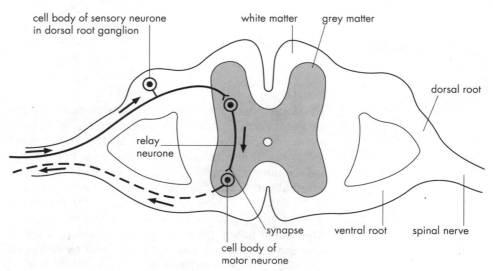

Fig. 10.5 Transverse section of the spinal cord

Function of the spinal cord

Its function is to relay information between the peripheral nervous system of the trunk and limbs and the brain. Most of the grey matter in the cord consists of **intermediate** or **association** fibres, which transmit impulses between sensory and motor fibres. These may be direct, as in the case of **spinal reflex arcs**, or may transmit to and from the brain.

PERIPHERAL NERVOUS SYSTEM

The afferent fibres bring impulses from the sense organs, **receptors**, to the central nervous system and efferent fibres take impulses to the **effector** organs. All the spinal nerves are mixed, carrying both sensory and motor fibres. The peripheral system forms a network throughout the body, supplying fibres to individual effector organ cells and receiving fibres from individual sensory cells. The basic unit of the peripheral nervous system and its association with the central nervous system is exemplified by a simple spinal reflex, as shown in Fig. 10.6. In practice, the situation is rarely as simple as this. Internuncial (intermediate) fibres are involved in the spinal reflex and they both relay information to the brain and carry impulses back from the brain. These impulses may modify the response; they go to the cell body of the motor neurone which integrates the information from the various sources to produce the response.

> 66 Usually large numbers of nerve cells are involved 99

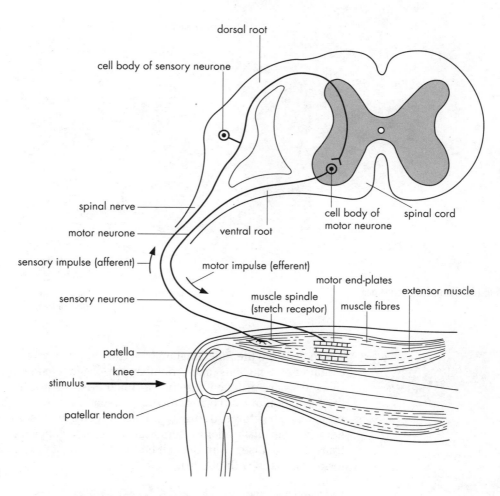

Fig. 10.6 Spinal reflex arc

AUTONOMIC NERVOUS SYSTEM

There are two sets of nerves forming this system, generally working antagonistically to each other to regulate the function of the internal organs.

Sympathetic system

The **sympathetic** system acts to increase the body's ability to be active, increasing the heart rate, dilating the bronchi and constricting blood vessels, so raising the blood pressure. Gut activity is reduced by the constriction of the sphincter muscles. This system consists of fibres which leave the spinal cord, forming **sympathetic ganglia** in pairs along the vertebral column. These join the spinal cord in each segment from the first thoracic to the fourth lumbar. Each ganglion is connected to the spinal cord by a **white ramus communicans** and to the spinal nerve by a **grey ramus communicans**. In some segments there are additional sympathetic ganglia near to the viscera. These are called **collateral ganglia** and may fuse together to form larger units: for example, the **solar plexus** in the abdomen. From these ganglia, post-ganglionic fibres run to the effector organs. The sympathetic effects are produced by the release of **noradrenalin** from the endings of the post-ganglionic fibres (see Fig. 10.7, p. 118).

Parasympathetic system

The **parasympathetic system** is antagonistic to the sympathetic, reducing the blood flow and respiratory rate and increasing the activity of the gut. These responses are brought about by the release of **acetylcholine** from the nerve endings. The parasympathetic system has ganglia close to the effector organs so the post-ganglionic fibres are very short. The long pre-ganglionic fibres originate in the brain and form part of the cranial nerves. Most of the parasympathetic fibres to the thorax and trunk are carried in cranial nerve 10, the **vagus**. Some parasympathetic fibres also originate in the sacral region of the spinal cord.

CHAPTER 10 CO-ORDINATION IN ANIMALS

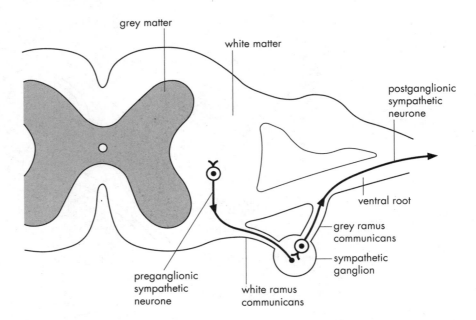

Fig. 10.7 Sympathetic nerves and ganglia

STRUCTURE OF NEURONES

These cells are the functional units of the nervous system. Each consists of a cell body containing a nucleus and granular cytoplasm containing many ribosomes. These ribosomes are grouped together forming **Nissl granules**, which are concerned with the formation of **neurotransmitter** substances. From each nerve cell body arise fine branching structures, **dendrites**. These receive impulses from other nerve cells. Some neurones also have a long, membrane-covered cytoplasmic extension, the **axon**, which transmits impulses from the cell body. At its end, an axon divides into branches which form **synapses** with other neurones.

Peripheral neurones

Peripheral neurones are surrounded by and supported by **Schwann cells**. In some cases, these grow round the axons of the nerve cells to form a multilayered **myelin sheath**. This acts as an electrical insulator and speeds up the transmission of impulses. Myelin sheaths are found only in vertebrate nervous systems. The myelin sheath has thin areas at intervals, **nodes of Ranvier**, which are important in impulse transmission. The structure of a generalised neurone is shown in Fig. 10.8.

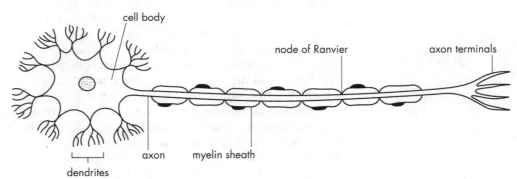

Fig. 10.8 Structure of a neurone

TRANSMISSION OF NERVE IMPULSES

The composition of the cytoplasm of an axon (**axoplasm**) is very different from that of the surrounding fluid. There is a great excess of potassium (K^+) ions in the axoplasm and an excess of sodium (Na^+) ions in the tissue fluid. This difference is maintained by the active transport of ions against their electrochemical gradient by a sodium/potassium-coupled cation pump. The net result of this ion imbalance is that the outside of the axon membrane is positively charged relative to the inside. To maintain this difference requires energy from ATP, which drives the cation pump (see Fig. 10.9). This is why a neurone contains many mitochondria and is metabolically very active. The ionic difference across the membrane produces an electrical potential of –60mV, which is known as the **resting potential**. Stimuli which alter this resting potential

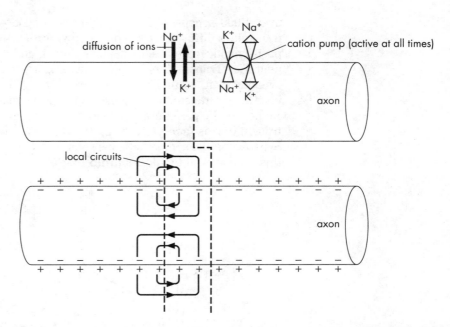

Fig. 10.9 Transmission of an impulse along a nerve fibre

sufficiently initiate nerve impulses along the axon; the polarisation of the axon is reversed to produce an **action potential**. This is an all or nothing response: if the stimulus is too small, a nerve impulse is not produced. The action potential causes a small electric current across the membrane and as a portion of the membrane is depolarised, depolarisation of the next portion is initiated. There is a series of local currents propagated along the axon. The sodium pump is active all the time and behind the transmission; this pump restores the resting potential. Once the resting potential is restored, another impulse can be transmitted. During the period between depolarisation and repolarisation impulses cannot pass. This is known as the **refractory period**.

> Be sure you understand the structural and functional differences between myelinated and unmyelinated fibres

When axons have a myelin sheath there is no contact between the axon membrane and the tissue fluid except at the nodes of Ranvier. The result is that the impulse jumps from one node to the next, speeding the overall passage along the axon. In mammals myelinated fibres transmit impulses at speeds of up to 100m s^{-1} which is 50 to 100 times the speed of unmyelinated fibres. Also the larger the diameter of the axon the greater the velocity of transmission (see Fig. 10.10).

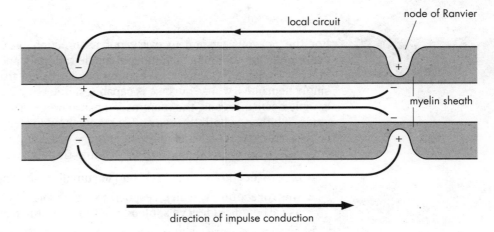

Fig. 10.10 Transmission of an impulse along a myelinated nerve fibre

SYNAPSES

Most junctions between neurones take the form of **chemical synapses**. Branches of axons lie close to dendrites of other neurones but do not touch; there is a gap of about 20μm between them. When impulses are transmitted, this gap is crossed by the secretion of a **neurotransmitter** from the axon membrane (**pre-synaptic membrane**), which diffuses across the space to stimulate the dendritic membrane (**post-synaptic membrane**). Although standard diagrams show nerve cells with few dendrites and therefore few synapses, the number of synapses with a single nerve cell can be very large. For example, a single spinal cord motor neurone may be involved in over 1,000 synapses. The structure of a synapse is shown in Fig. 10.11. When the transmitter diffuses across the gap (**synaptic cleft**) it attaches to a receptor site on the post-synaptic membrane, depolarising it and so initiating an impulse in the next neurone. Transmitters, when released, are quickly destroyed by enzymes in the synaptic cleft, so their effect is limited and the merging of impulses is prevented. If insufficient transmitter is released, the post-synaptic membrane will not be stimulated. In most of the peripheral nervous system synapses and at neuromuscular junctions the transmitter is **acetylcholine**, the enzyme destroying it being **cholinesterase**. Synapses in the sympathetic nervous system use **noradrenalin** as the transmitter and this is destroyed by **monoamino oxidases**.

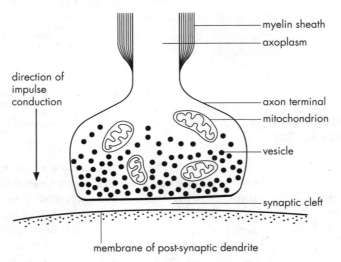

Fig. 10.11 A synapse

TYPES OF SYNAPSE

The are two types of synapse.

> This distinction is important in neural functioning

- **Excitatory synapses** The neurotransmitter opens sodium and potassium channels in the post-synaptic membrane, leading to its depolarisation.
- **Inhibitory synapses** The transmitter opens potassium and chloride channels, causing the membrane to become **hyperpolarised** and so unable to transmit the impulse.

SENSORY RECEPTORS

For the nervous system to carry out its function properly it is dependent upon a continuous input of information from inside the body and from the environment. This input is initiated by sensory receptors, which range from specialised sensory cells, single neurones, whose surface is capable of detecting a stimulus, through to the most complex sense organs, the mammalian ear and eye. Most A-level syllabuses include a study of either or both the eye and ear. You should refer to the standard text books for details of the structure and function of these organs.

Sensory receptors can be subdivided into three groups:

- **exteroreceptors**, which respond to stimuli originating in the external environment
- **interoreceptors**, which respond to physiological changes within the body, such as carbon dioxide levels, blood pressure and the osmotic pressure of the blood
- **proprioceptors**, which receive stimuli from the muscles and skeleton and are concerned with muscle movement and body position

Alternatively, receptors can be classified according to the type of stimulus received:

- **photoreceptors** – light
- **mechanoreceptors** – sound, touch, pressure, gravity
- **chemoreceptors** – taste, smell
- **thermoreceptors** – temperature change

Fig. 10.12 shows some of these.

CHAPTER 10 ESSENTIAL PRINCIPLES

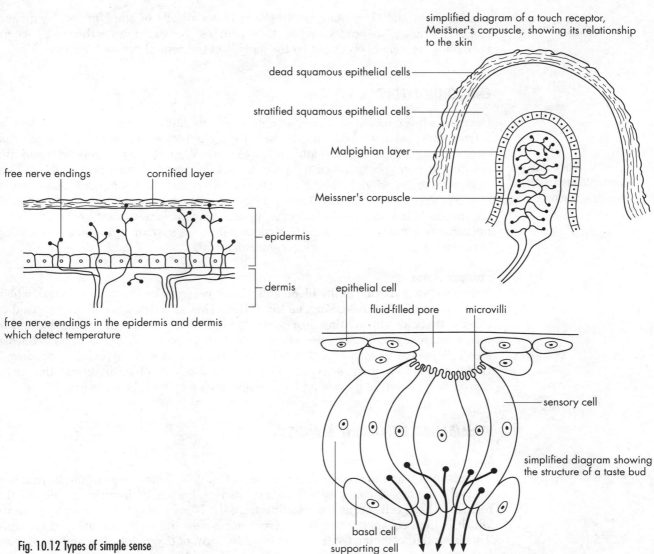

Fig. 10.12 Types of simple sense organ

MODE OF ACTION

66 Transduction is an important concept **99**

Sensory receptors are detecting one form of energy and converting it into electrical energy: that is, nerve impulses. They are acting as **transducers**. Sensory cells at rest have a polarised membrane like nerve cells. When stimulated, the membrane depolarises, producing a **generator potential**. The magnitude of the generator potential varies with the strength of the stimulus. When this reaches the threshold, an action potential is produced in the sensory axon leaving the sensory cell. The mechanism is the exchange of sodium and potassium ions, as in a neurone.

- In chemoreceptors the depolarisation is produced by the relevant chemicals attaching to receptor sites or entering the membrane.
- In photoreceptors light penetrates the cells and produces the generator potential by the chemical changes involved when visual pigments are split.
- In thermoreceptors temperature changes affect the membranes.
- Mechanoreceptors are stimulated by distortion of the membrane.

ENDOCRINE GLANDS AND THEIR SECRETIONS

The endocrine system is a system of ductless glands distributed throughout the body of a mammal. Hormones are secreted directly into the bloodstream and are transported by the blood to their target organs. This distinguishes endocrine organs, from exocrine organs which have ducts to channel their secretions directly to their place of action. You should refer to Chapter 7, Homeostasis, and Chapter 12,

Reproduction and Development in Mammals, for details of some of the hormones. Hormones are also produced by other animals: for example, arthropods, where hormones are mainly produced by the ganglia of the central nervous system.

ENDOCRINE FEEDBACK SYSTEM

Hormones are concerned with the control of body function, regulating the levels of activity in relation to body needs and playing a part in maintaining the internal environment within narrow limits. These control mechanisms depend upon the constituent parts being linked together so that the activity of one part or its product influences the activity of other parts. This system is known as **feedback** and it works to increase or decrease secretions in relation to the optimum level of the function or constituent being controlled. Endocrine system function is controlled by the process of **negative feedback**, which increases the stability of a system. There is a relationship between the product and the releasing mechanism.

> *Make sure you understand the concept of negative feedback*

Example: thyroxine

If the level of thyroxine in the blood falls, this is detected by the hypothalamus, which produces **thyroxine releasing factor (TRF)**. This stimulates the pituitary gland to release **thyroid stimulating hormone (TSH)**, which causes the thyroid gland to release more thyroxine. The rise in thyroxine level causes the hypothalamus to release less TRF, which in turn cuts down TSH release and so less thyroxine is produced. Under normal circumstances this negative feedback mechanism keeps the blood thyroxine level within narrow limits. The feedback loop was shown in Fig. 7.7.

MAMMALIAN ENDOCRINE GLANDS

The thyroid gland

The gland is found in the neck, in the form of flat lobes either side of the larynx. It is formed of cuboidal epithelial cells which secrete hormones into the cavities of the follicles. These cells accumulate **iodine** from the blood and combine it with **tyrosine**, which is then converted into two thyroid hormones, the main one being **thyroxine**. This is stored, combined with a protein in the form of **thyroglobulin**, and is released into the blood under the influence of TSH. Thyroid hormones regulate growth and development by influencing cell growth and differentiation. They also control the basal metabolic rate.

- **Undersecretion** in children gives rise to **cretinism** unless thyroxine is administered. In adults undersecretion causes **myxoedema**, where the basal metabolic rate is too low and too little energy is released, with a consequent increase in body mass.
- **Oversecretion**, **hyperthyroidism**, gives rise to overactivity and consequent weight loss.

The pancreas

The endocrine portion is the **islets of Langerhans**. These contain two types of secretory cells, known as α and β cells, which secrete **glucagon** and **insulin** respectively. These two hormones work antagonistically to control the blood sugar (glucose) level. The relationship is shown in Fig. 10.13. If the pancreas fails to produce sufficient insulin, the blood glucose level rises and excess glucose is excreted by the kidneys. This condition is **diabetes mellitus** and if not treated, either by administering insulin or by regulating the diet, it can lead to serious illness or death.

The adrenal gland

Anatomically this consists of two separate parts, an inner **medulla** and an outer **cortex**. The main secretion of the medulla is **adrenalin** but a little **noradrenalin** is also secreted. These hormones are constantly secreted but when a mammal is threatened and needs to fight or escape from an enemy, the level of secretion is greatly increased. These secretions also increase whenever an animal is subject to psychological or mental stress. The overall effects of these hormones are:

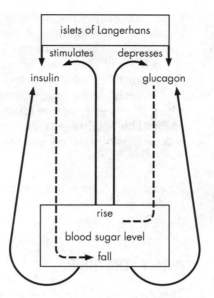

Fig. 10.13. Hormonal control of blood glucose levels

- to reduce the activity of the gut by relaxing the smooth muscle
- to increase the heart rate, cardiac output and blood pressure
- to raise the blood sugar level, enchancing the animal's ability to perform physical activity

The cortex secretes two sets of hormones:

- **Glucocorticoids**, e.g. **cortisol**, raise the blood sugar level by stimulating the conversion of fats and proteins into glucose; they are also important in controlling inflammation and promoting repair of body tissues.
- **Mineralocorticoids** are the other group of hormones; the most important one is **aldosterone**, which is discussed in Chapter 7.

INSECT HORMONES

Hormones in insects are involved in the control of metamorphosis. The hormones are secreted by neurosecretory cells in the brain. These hormones pass along nerve axons to two pairs of bodies, the **corpora allata** and **corpora cardiaca**, which lie behind the brain. The other endocrine glands are the **prothoracic glands** in the first thoracic segment (see Fig. 10.14). The corpora allata store and secrete **juvenile hormone**,

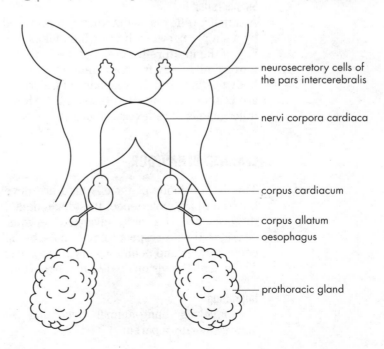

Fig. 10.14 Insect endocrine glands

which suppresses the development of pupal or adult characters, thus allowing larval growth and development. A trophic hormone is produced by the neurosecretory cells and stored in the corpora cardiaca. This passes to the prothoracic glands, stimulating the release of **ecdysone** or moulting hormone. This initiates moulting and also the development of adult structures. Both hormones are needed for larval or nymphal development; before each moult there is a reduction in the level of juvenile hormone, which allows ecdysone to produce the moult. The level of juvenile hormone then rises again. The level of juvenile hormone is permanently reduced to allow pupation in holometabolous insects and the emergence of the adult in all insects.

ANIMAL BEHAVIOUR

Animal behaviour is the result of the co-ordination systems receiving information and producing responses to it. Although the distinction is not always clear cut, for descriptive purposes behaviour can be divided into two forms: **innate** and **learned**.

INNATE BEHAVIOUR

> Do not use the term instinctive behaviour

This includes all the behavioural responses which are predetermined by the inherited response pathways and are concerned with the survival of the organism. Some of the patterns are very simple, such as stopping or avoidance reactions, but others, such as web-building in spiders, are very complex. Three types of simple innate behaviour feature in A-level syllabuses.

Taxis
A taxis is the movement of a whole organism in response to a directional stimulus. It may be towards (positive), away from (negative) or at an angle to, a stimulus. In many cases, for example, planaria or blowfly larvae, the stimulus direction is perceived by moving the head from side to side, allowing paired receptors to detect the direction by the varying stimulation of each. The animal orientates itself until the stimulus on both is equal. Many different stimuli, for example, light, temperature or chemicals, can produce tactic responses.

Kinesis
A kinesis is a nondirectional movement response in which the intensity, not the direction of the stimulus, is important. In a kinetic response the rate of movement changes with increased intensity of stimulus. For example, the rate of movement of the tentacles of *Hydra* increases in the presence of food and the rate of locomotion of woodlice increases in dry conditions.

Simple reflex
A simplex reflex is an automatic stereotyped response to a given stimulus, determined by a simple nerve pathway which takes the form of a **reflex arc** and does not involve any of the interpretive areas of the brain. Simple reflexes include flexion responses to move limbs away from a painful stimulus, stretch responses, involved in maintaining body posture, and the response of the iris to light intensity. The simplest form of reflex arc is shown in Fig. 10.6 (see p.117). However, you should refer to standard texts for fuller details of reflex arc structure.

LEARNED BEHAVIOUR

This is behaviour patterns which are developed during the lifetime of an organism based on past experience. It is dependent on memory storing experience gained and the presence of a fairly advanced nervous organisation. Learned behaviour is often considered with respect to mammals, but all animal groups except protozoa, sponges, coelenterates and echinoderms show learned behaviour to a greater or lesser extent. The main subdivisions of learned behaviour are:

Imprinting
This involves young animals becoming associated with and identifying with another animal, usually a parent.

Classical conditioning
This develops where animals learn to produce a conditioned response to a conditioned stimulus. This is also known as a **conditioned reflex** and is exemplified by the experiments of Pavlov with dogs.

Operant conditioning
Operant conditioning or trial and error learning occurs where the responses are reinforced by reward or punishment, as shown by Skinner's work with pigeons.

Habituation
This occurs when continuous repetition of a stimulus not associated with a reward or punishment leads to the disappearance of a response.

Exploratory learning
This is not associated with immediate needs but allows the animal to store information which might be useful later.

Insight learning
This is based on thought and reasoning, using past experience. It is thus limited to those animals with a well-developed cerebrum and is found mainly amongst the primates.

SOCIAL BEHAVIOUR

Many species of insects and many vertebrates show social behaviour, where groups of animals live together and co-operate. One of the bases for a social group is a system of communication. This system may be:

- **chemical** – insects
- **visual**, via body language, facial expression, etc. – primate species
- mainly **vocal** – dolphins and humans

Usually there is some form of social organisation. This can take the form of:

- a **caste system** – bees, ants, termites
- a **social hierarchy** – poultry, pack carnivores, baboons

Social organisations always have survival value, which can mean such things as increased protection against predators, enhanced food-gathering, improved hunting, protection against adverse weather conditions or increased overall efficiency by a division of labour among group members.

PRACTICAL WORK

The structure of the spinal cord can be studied by the microscopic examination of sections of a spinal cord. The anatomy and major functional areas of the brain should be studied by looking at the surface and sagittal sections of preserved brains. The transmission of nerve impulses is not easy to study practically at A-level but there are videotapes and computer programs available which allow work on the recording of the electrical changes involved and also the electrochemistry of nerve impulses.

The function of sense organs can be investigated using your own eyes, ears, taste, smell and skin senses. The various practical guides available give details of suitable and safe experiments.

Behaviour experiments to demonstrate kineses and taxes can be carried out using small invertebrates such as woodlice or *Tribolium* in choice chambers. Taxes can also be demonstrated using blowfly larvae and unilateral light. An investigation of short-term memory, learning and reflex action can be carried out in groups with your fellow students.

Further reading
Boyce A and Jenkin M, 1986, *Metabolism, Movement and Control*. Macmillan
Buckle J W, 1983, *Animal Hormones*. Arnold
Jones D G, 1981, *Neurones and Synapses*. Arnold
Reiss M and Sants H, 1987, *Behaviour and Social Organisation*. CUP

CHAPTER 10 CO-ORDINATION IN ANIMALS

EXAMINATION QUESTIONS AND ANSWERS

QUESTIONS

Question 10.1

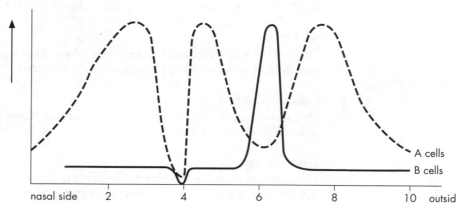

Fig. 10.15

The graph shows the number of receptor cells (Types A and B) in the human retina along a horizontal line from the nasal side of the eye to the outer side. The distances are expressed in arbitrary units.
a) Identify the receptor cells A and B. (2)
b) Using the diagram, describe and comment on the distribution of the cells in the human retina. (5)

(Total 7 marks)
(ODLE)

Question 10.2
The following actions may be stimulated by hormones in a mammal. Write the name of *one* appropriate hormone beside each of the actions described.

ACTION	HORMONE
Ejection of milk from mammary glands	
Hydrolysis of glycogen in liver	
Reabsorption of sodium ions in kidney	
Ovulation	
Acceleration of heart rate	
Development of male secondary sexual characteristics	
Contraction of smooth muscle in the uterus wall	
Increased glucose uptake by muscle cells	

(8)
(Total 8 marks)
(ULEAC)

Question 10.3
The diagram below shows part of two nerve fibres and a synapse. The figures indicate the value in mV of the potential across the membrane between the cytoplasm of the fibres and the extracellular fluid at intervals along each fibre.

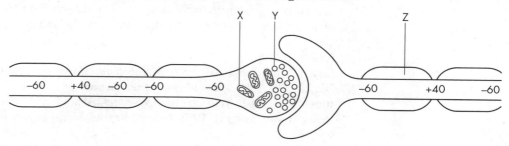

Fig. 10.16

a) i) Draw a circle round one region of the diagram where an action potential exists. Explain your choice. *(2)*
   ii) By means of an arrow on the diagram, indicate the direction in which action potentials would normally travel along these fibres. *(2)*
b) Identify structures X and Y and state how each is involved in the transmission of nerve impulses. *(4)*
c) i) What is the major chemical constituent of structure Z? *(1)*
   ii) State two effects of structure Z on the transmission of action potentials. *(2)*

(Total 11 marks)
(ULEAC)

Question 10.4

Termites are small wingless insects found in large colonies in the soil in many dry areas. At certain times in the year, winged males and females are produced which mate and produce new colonies.

In an experiment investigating behavioural responses, untreated termites treated in various ways were released on a sloping surface illuminated from one side. Traces A,

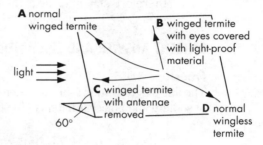

Fig 10.17

B, C and D in the diagram show typical results.
a) i) To what stimuli is the normal winged termite (trace A) responding?
   ii) What type of response is shown by termites in these traces? *(2)*
b) Explain the response of the termites in:
   i) trace B; ii) trace C. *(4)*
c) Explain the importance of trace A and trace D in the life of the termite. *(2)*

(Total 8 marks)
(AEB)

OUTLINE ANSWERS

Answer 10.1
a) A are rods; B are cones
b) Number of cells A and B varies at different points across the retina; none of A or B at the point 4 units from the nasal side; this is the blind spot; many of type B but few of type A at 6 units from the nasal side; this is the fovea – many B cells give high acuity and colour vision; many more A cells at periphery; give poor light vision but no colour vision; no receptors at 0 or 10 units as little light received here.

Answer 10.2
prolactin/oxytocin
glucagon
aldosterone
FSH
adrenalin
testosterone
oxytocin
insulin

Answer 10.3
a) i) Circle round a '+40' point; polarisation is reversed at this point and the membrane is depolarised.

ii) Arrow left to right; the impulse crosses a synapse from the terminal button to the post-synaptic membrane.
b) X = mitochondrion, produces ATP/produces energy.
Y = secretory vesicle, releases neurotransmitter/acetylcholine.
c) i) Phospholipid/fatty material/myelin
ii) Increases resistance to current flow between the axoplasm and the extra-cellular fluid; increases the speed of transmission/produces saltatory conduction.

Answer 10.4
a) i) Light and gravity   ii) Taxis
b) i) There is now no response to light. The winged termite shows negative geotaxis and so moves upwards.
ii) There is now no response to gravity. The winged termite shows positive phototaxis.
c) i) The winged termites will move away from gravity and towards light. They will thus move out of the colony to form new ones.
ii) Wingless termites will move away from light and towards gravity. They will therefore stay in or near the colony.

STUDENT ANSWER AND EXAMINER COMMENTS

Question 10.5
The line diagram shows an axon membrane. The table to the right gives the concentrations of sodium and potassium ions in the fluids inside and outside the axon membrane.

ION	CONCENTRATION IN MOL kg^{-1}	
	OUTSIDE	INSIDE
K$^+$	20	400
Na$^+$	460	50

a) The different concentrations of ions results in a potential difference between the inside and outside of the axon. Which is positive and which is negative? Explain your answer.

"Both correct, only one need be stated"

Outside the axon is positive.
Inside the axon is negative.

"Correct, but more detail than required, 'higher concentration of positive ions outside' is sufficient"

Both potassium and sodium ions are positively charged with a single charge. From the table the total concentration of ions outside the axon is 480 and inside is 450.
Therefore, relatively, the charge is more positive outside than inside.

b) The different ion concentrations are brought about mainly by a Na$^+$/K$^+$ coupled cation pump, which actively exchanges ions across the membrane. Copy and complete the diagram by adding arrows to show the movements of ions caused by the pump. *(1)*

"Correct"

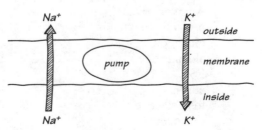

Fig 10.18

c) What provides the energy to drive the pump?

"Correct"

The energy to drive the pump comes from the hydrolysis of ATP. *(1)*

(Total 4 marks)
(ODLE)

REVIEW SHEET

1. Briefly describe (in each case using a diagram) an experiment to demonstrate each of the following.
   a) That the growth response to light is just behind the tip of an oat coleoptile.

   b) That there is some growth-promoting substance which diffuses away from the illuminated side to the darkened side.

2. Define and briefly explain the factors involved in **geotropic response**.

3. Explain the role played by **gibberellin** in germination.

4. Define each of the following, giving examples wherever appropriate.
   a) Nastic movements:

   b) Tactic movements:

5. Explain the roles of **phytochrome** in plants.

6 Complete the labels on the diagram of the earthworm. Then use your diagram to explain the operation of the nervous system of the earthworm.

_____
_____
_____
_____
_____
_____
_____
_____
_____
_____

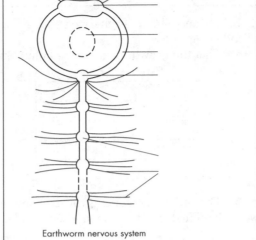

Earthworm nervous system

7 Consider the structure and function of the **forebrain** in mammals.

_____
_____

8 Complete the labels on the following diagram. Then use the diagram to explain the workings of the **sympathetic nervous system**.

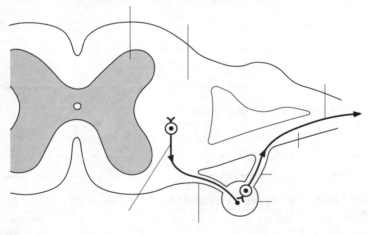

_____
_____

9 Present an appropriate diagram and use it to explain the **transmission of nerve impulses**.

_____
_____

10 Describe the structural and functional differences between **myelinated** and **unmyelinated** fibres in the transmission of nerve impulses.

_____
_____

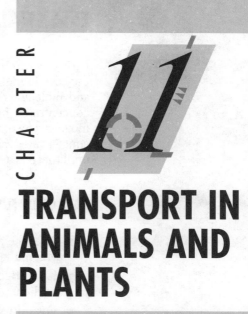

CHAPTER 11
TRANSPORT IN ANIMALS AND PLANTS

DIFFUSION

FACILITATED DIFFUSION

ACTIVE TRANSPORT

TRANSPORT SYSTEMS IN ANIMALS

BLOOD

TRANSPORT IN PLANTS

PRACTICAL WORK

GETTING STARTED

Chapters 5 and 6, dealing with the acquisition of food materials by living organisms, and Chapter 8, on respiration and gas exchange, lead up to this next topic, transport, as both the food and the respiratory gases must be transported around the body of the organism. An understanding of these previous chapters would help before tackling this new topic. It would also help to refer back to Chapter 4, on Cell Biology, to remind yourself of the kinds of molecules which are being transported and the nature of the cell membrane.

The examination syllabuses show a remarkable degree of agreement in the subjects included in this section, so you should not find that yours differs markedly from the contents of this chapter, but of course it is always wise to check.

Before we look at animals and plants as whole organisms and consider their transport mechanisms, we must look at the ways in which materials get into and out of cells. The following processes will be described:

- diffusion
- facilitated diffusion
- active transport
- pinocytosis
- phagocytosis

CHAPTER 11 TRANSPORT IN ANIMALS AND PLANTS

ESSENTIAL PRINCIPLES

DIFFUSION

Diffusion can be defined as the movement of **molecules** or **ions** from a region where they are in high concentration to a region where they are in lower concentration. Ions and molecules are always in a state of **random movement**, but if they are highly concentrated in one area, there is a greater probability of them moving away than moving closer together. There will be a net movement away from the highly concentrated region. The movement away will occur until equilibrium is reached, or until there is a uniform distribution of ions or molecules.

RATE OF DIFFUSION

- The rate at which diffusion occurs is in inverse proportion to the distance over which it takes place, so it is most effective over short distances.
- It is affected by the **concentration gradient**; the steeper the concentration gradient, the faster the rate of diffusion.
- For diffusion to occur in the bodies of organisms, the cell membranes must be **permeable** to molecules and ions. All cell membranes are permeable to oxygen, carbon dioxide and water.

Diffusion accounts for:

- the movement of oxygen into cells for respiration
- the movement of carbon dioxide out of cells as a waste product
- the distribution of food materials

FACILITATED DIFFUSION

Some molecules, such as oxygen and carbon dioxide, diffuse rapidly through the cell membrane in solution, but ions and small polar molecules diffuse through much more slowly because they are insoluble in the lipid part of the membrane. Facilitated diffusion is a modified form of diffusion in which molecules and ions are transported across the cell membrane by **protein carriers**. The molecules are picked up on one side of the membrane, where they are in high concentration, and then deposited on the other side of the membrane, where they are in lower concentration. This type of diffusion is shown by the movement of glucose into red blood cells (see Fig. 11.1). Facilitated diffusion does not involve the expenditure of energy and does not take place against the concentration gradient.

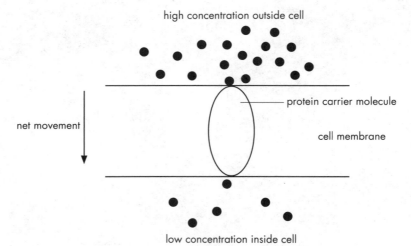

Fig. 11.1 Facilitated diffusion

OSMOSIS

Osmosis is defined as the passage of solvent molecules from a region where they are in high concentration to a region where they are in lower concentration through a partially permeable membrane. In biological systems the **solvent** is **water**, and the

partially permeable membrane is the **cell membrane**, through which some molecules can diffuse readily while others penetrate more slowly or not at all.

Demonstration of osmosis

> Be familiar with this simple demonstration

A simple demonstration of this can be made using an **osmometer**, consisting of an inverted thistle funnel, the mouth of which has been covered by Visking tubing (a partially permeable membrane) and filled with a concentrated sugar solution. The thistle funnel is lowered into a beaker of water, the level of the sugar solution is marked and then the apparatus is left for some time. Later it can be seen that the level of the sugar solution has risen up the thistle funnel. This can be explained by the net movement of the water (solvent) molecules through the membrane from where they are in high concentration (the beaker), but the sugar molecules (solute) are unable to move in the opposite direction because of the nature of the membrane.

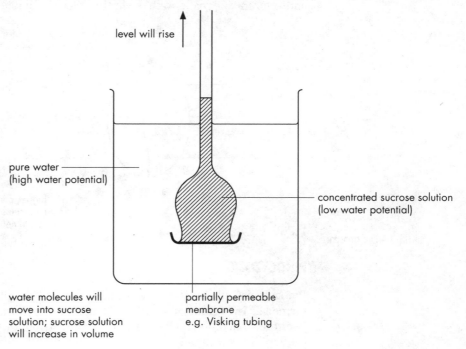

Fig. 11.2 An osmometer

Osmotic pressure

In a situation where pure water is separated from a sugar solution by a partially permeable membrane, the **hydrostatic pressure** needed to prevent water molecules passing through that membrane into the solution is called the osmotic pressure of that solution. The more concentrated the sugar solution, the higher its osmotic pressure will be, but it can be measured as a real pressure only when it is in an osmometer.

Osmotic potential

Normally the pressure is only a potential pressure and is referred to as the osmotic potential, with a negative sign. It has been the practice for some time to express the concentration of a solution in terms of its osmotic pressure but, as has been explained, it is more accurate to refer to osmotic potential.

Water potential

The term water potential (indicated by the Greek letter psi ψ) is now currently used by biologists to describe the tendency of water molecules to move from high to low concentrations: water will diffuse from a region of high water potential to a region of lower water potential.

- Where there is a high concentration of water molecules, the water molecules have a greater potential energy; a higher water potential implies a greater tendency to leave.
- Where there is a lower concentration of water molecules, as in a concentrated solution, the potential energy of the water molecules is less; a lower water potential implies less tendency to leave.

The water potential of pure water at atmospheric pressure is arbitrarily given a value of zero, so all solutions have lower potentials than pure water and negative values.

ACTIVE TRANSPORT

Active transport is an energy-requiring process in which ions and molecules are moved across membranes **against concentration gradients**. Thus ions and molecules are being moved in the opposite direction to that in which diffusion occurs. The energy is supplied by the **hydrolysis of ATP**, and so anything which interferes with the respiratory process will affect active transport. It takes place by means of **carriers**: protein molecules which span the cell membrane. The molecules to be transported become attached to the protein molecules and are then moved into the cell, probably by means of a change in shape of the carrier molecule (see Fig. 11.3). There are many cases where the movement of ions and molecules is brought about by active transport, and reference is made to these in the relevant chapters.

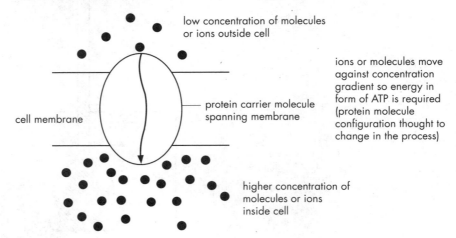

Fig. 11.3 Active transport

PINOCYTOSIS

Pinocytosis is the way in which liquids can be brought into cells through tiny channels formed as **invaginations** of the cell membrane. From the ends of these channels small **vacuoles** become pinched off, enabling distribution and more efficient absorption of the liquid taken up. The vacuoles are lined with membrane and the contents do not become incorporated into the cell until they have passed through it, so this is a method of speeding up the process of uptake.

PHAGOCYTOSIS

Phagocytosis provides a means whereby large particles can be taken up by cells. It can be carried out only by specialised cells, such as the **granulocytes** in the blood and by *Amoeba*. The particle to be taken up is surrounded by an invagination of the cell membrane and eventually enclosed in a vacuole. The particle is digested by enzymes, secreted into the vacuole by **lysosomes**, and the soluble products of digestion diffuse into the surrounding cytoplasm.

- **Exocytosis** refers to the emptying of membrane-lined vacuoles at the cell surface.
- **Endocytosis** is used to describe the uptake of materials into vacuoles.

TRANSPORT SYSTEMS IN ANIMALS

The actual exchange of materials at the cellular level takes place by a variety of different means, which have been discussed already, but these mechanisms, efficient though they may be over short distances, would not be rapid enough to enable the distribution of essential metabolites to all parts of a complex multicellular organism. In unicellular organisms and those multicellular organisms with a large surface area to volume ratio, transport by diffusion seems to be efficient and special transport systems unnecessary. In the larger organisms, because of their bulk, there are specialised areas for gas exchange and for the absorption of food; a transport system provides the link between these areas and the cells which require oxygen and nutrients.

Simple organisms, such as the **Protozoa**, do not require a specialised transport system as their surface area to volume ratio is such that they can obtain soluble food materials by diffusion from the vacuoles, and obtain oxygen for respiration and get rid of their waste products by diffusion through the plasma membrane.

BLOOD TRANSPORT SYSTEMS

Transport systems involving **blood** are found in all the vertebrates and in many invertebrates as well. Blood systems usually consist of tubular vessels through which blood is circulated by means of a pumping device.

There are two types of **vascular** system:

- the **open** vascular system, in which the blood is pumped by the heart into a series of blood spaces called **haemocoel**
- the **closed** vascular system, in which the blood is confined to the blood vessels

> Insects have open vascular systems

Example of an open system
The circulatory system in **insects** is of the open type. The heart is a tubular vessel which pumps blood into the haemocoel, where it bathes the body organs. Eventually the blood gets back into the heart to be pumped out again, but the distribution of blood is poorly controlled. It is important to remember here that the blood of insects is not involved in the transport of respiratory gases.

> Vertebrates have closed vascular systems

Example of a closed system
The circulatory system in the **earthworm** is of the closed type. The pumping device consists of five pairs of contractile '**pseudohearts**' which control the flow of blood and link the dorsal blood vessel with the ventral vessel. There is haemoglobin in the blood of the earthworm but it is dissolved in the plasma.

CIRCULATION

All the vertebrates have closed circulatory systems, with **muscular hearts**, but there are differences in the routes taken by the blood.

- **Single circulation** This is evident in **fish**, where the blood is pumped by the heart to the gills to be oxygenated; from there it goes to the rest of the body and then back to the heart.
- **Double circulation** This is evident in **mammals**, where the heart is completely divided into two parts; blood is pumped from the heart to the lungs and back to the heart in one circulation, and then pumped to the body in a completely separate circulation. Therefore, for a complete circuit of the body, the blood passes through the heart twice.

BLOOD

Blood in the vertebrates is a fluid made up of **cells** in **plasma**.

PLASMA

The plasma is composed largely of water, over 90%, with soluble food molecules, waste products, hormones, plasma proteins, mineral ions and vitamins dissolved in it.

CELLS

The cells are of three types:

> Types of blood cell

- **erythrocytes** – red blood corpuscles
- **leucocytes** – white blood corpuscles
- **thrombocytes** – platelets

Erythrocytes

The function of the erythrocytes is the carriage of oxygen from the respiratory surface to the respiring cells; they are highly adapted for this function. Mammalian erythrocytes do not contain a nucleus, so have a biconcave shape. They are filled with the red pigment **haemoglobin**, a single corpuscle containing as many as 250 million molecules. Their function is explained in more detail in the chapter on respiration and gas exchange, Chapter 8.

Leucocytes

There are two groups of leucocytes:

- **granulocytes**, which are **phagocytic**, have granular cytoplasm, lobed nuclei and engulf bacteria
- **agranulocytes**, which produce **antibodies** and **antitoxins**, have clear cytoplasm and spherical or bean-shaped nuclei

> **Check with your syllabus**

Some syllabuses do require a more detailed knowledge of the different types of leucocytes, especially if there are options where reference is made to the immune system.

Thrombocytes

The thrombocytes are irregularly shaped fragments of cells which are involved with the clotting mechanism of the blood. If they are damaged or exposed to air, they release **thromboplastin**, which helps to convert **prothrombin** to **thrombin**, an enzyme catalysing the conversion of the soluble plasma protein **fibrinogen** to insoluble threads of **fibrin**, which helps to seal the wound.

BLOOD VESSELS AND THE CIRCULATION

Blood is pumped by the heart into thick-walled vessels called **arteries**. These split up into smaller vessels called **arterioles**, from which the blood passes into the **capillaries**. The capillaries form a vast network which penetrates all the tissues and organs of the body. Blood from the capillaries collects into **venules**, which in turn empty blood into **veins**, from where it is returned to the heart.

Structure of arteries and veins

The arteries and the veins have the same basic three-layered structure (see Fig. 11.4), but the proportions of the different layers do vary. In both:

- the **innermost layer** is the **endothelium**, which is one cell thick and provides a smooth lining with minimum resistance to the flow of the blood
- the **middle layer** is made up of **elastic fibres** and **smooth muscle**, and this is the layer which is thicker in the arteries than in the veins
- the **outer layer** is made up of **collagen fibres**, which are resistant to overstretching

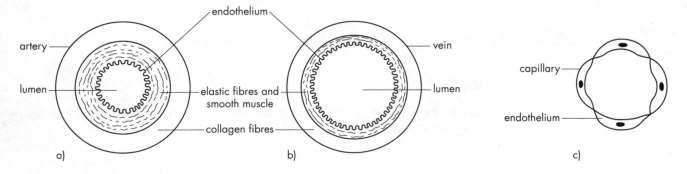

Fig. 11.4 Structure of a) an artery; b) a vein; and c) a capillary

Veins have **watch pocket** or **semi-lunar valves** in them; these are not present in arteries. The capillaries are 7 to 10 μm in diameter and consist only of a layer of endothelium, so their walls are permeable to water and dissolved substances such as glucose. It is at the capillaries that the exchange of materials between the blood and the tissues take place.

The heart

The heart is situated in the thorax between the two lungs. It is four-chambered, with **right** and **left atrium** and a **right** and **left ventricle**. It is a requirement of some syllabuses that candidates be able to dissect a mammalian heart, and some others include the heart in the dissection of the thorax. It would be wise to check your syllabus to make sure that you are familiar with the correct dissection.

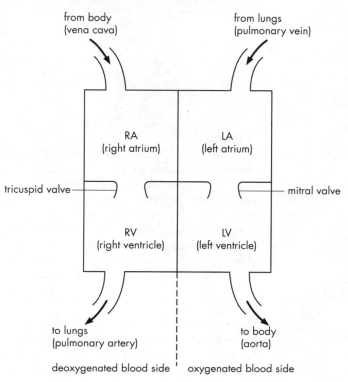

Fig. 11.5 Simplified diagram of the heart

Blood flow

Blood enters the right side of the heart from the **venae cavae** and passes into the thin-walled right atrium, from where it passes into the thicker-walled right ventricle, partly by gravity and partly by contraction of the atrial wall. It passes through the **atrio-ventricular valve**, called the **tricuspid** valve (three flaps), and is then forced into the **pulmonary arteries** when the ventricle muscle contracts. When this happens the tricuspid valve is forced to close, preventing back-flow of blood into the atrium, and the **semi-lunar valves** in the walls of the **pulmonary arteries** are forced open. When the ventricle muscle relaxes, the semi-lunar valves close, preventing back-flow of blood into the ventricle. More blood then comes into the ventricle from the atrium.

Events have been described for the right side of the heart, but both sides of the heart contract simultaneously, so the same is happening on the left-hand side. The only difference is that the ventricle wall on the left is much thicker and a greater force is generated when it contracts, forcing the blood into the aorta and then into general circulation. Reference should be made to the tendinous cords which attach the flaps of the atrio-ventricular valves to the papillary muscles in the walls of the ventricles. When the ventricle contracts the papillary muscles contract and pull on the tendinous cords, so pulling on the flaps and preventing them from turning inside out.

MECHANISM OF THE CONTRACTION OF THE HEART

Cardiac muscle

The heart can continue to contract rhythmically for a very long time without tiring, possible because of the nature of the **cardiac** muscle. This muscle consists of a network of interconnected fibres which show the same type of cross striations as skeletal muscle but are divided up into uninucleate cells. There is a rapid spread of excitation through the muscle due to the interconnections, resulting in a synchronised contraction. The cardiac muscle is **myogenic**: that is to say, its rhythmical contractions are generated within the muscle itself and are not due to nervous stimulation.

CHAPTER 11 TRANSPORT IN ANIMALS AND PLANTS

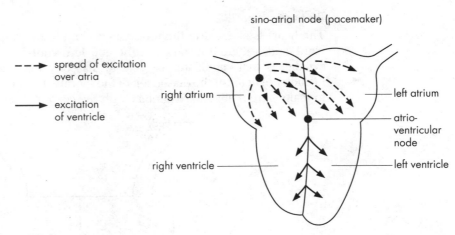

Fig. 11.6 The heart's pacemaker and its effect

Pacemaker
In the wall of the right atrium there is a region of specialised cardiac fibres called the **sino-atrial node (SAN)**, which acts as a **pacemaker** and initiates the heartbeat. A wave of electrical stimulation arises at this point and then spreads over the two atria. This causes the atria to contract. The stimulation reaches another specialised region of cardiac fibres, the **atrio-ventricular node (AVN)**, and causes that to pass on the excitation to specialised tissue in the ventricles. From the AVN, the excitation passes along the **bundle of His** and then spreads through the **Purkinje tissue** in the walls of the ventricles. As with the atria, stimulation is followed by contraction and the walls of the ventricle contract simultaneously.

Heart rate
The rate at which the heart beats can be modified by the nervous system; stimulation of the **vagus nerve** will slow down the heartbeat, but stimulation of the **sympathetic nerve** will accelerate it. The heartbeat can also be influenced by many non-nervous stimuli, such as temperature and hormones.

INTERCELLULAR FLUID AND LYMPH

In many syllabuses reference is made to the formation of intercellular fluid and its relationship with the blood and the lymph.

- Intercellular fluid is formed when blood passes through the capillaries and due to the blood pressure, some of the components of the plasma are forced out between the cells. It is from this tissue fluid that the cells get their supplies of oxygen and nutrients by diffusion, and the waste products of metabolism pass into it. Some of this fluid passes back into the capillaries, but some drains into the **lymphatic system** and is returned to the blood eventually, via the **thoracic duct**, which empties into the sub-clavian vein.

- Lymph is the tissue fluid that drains into blind-ending lymphatic capillaries amongst the tissues that join up to form larger vessels. It is moved by contractions of the muscles through which the vessels pass, being prevented from back-flow by numerous valves similar to those found in the veins. There are lymph **glands** and **nodes** associated with the lymph vessels, and these play an important role in the formation of lymphocytes and the prevention of infection.

Some syllabuses require a knowledge of the histology of cardiac muscle, and others require identification of specific blood vessels in the body, as seen in a dissection. The structure and composition of blood is a requirement for nearly all syllabuses, and even if it is not specified, it is necessary in order to understand the transport and protective functions.

TRANSPORT IN PLANTS

Transport in plants is principally involved in getting the materials for photosynthesis to the cells and then moving the products away to other parts of the plant. In simple organisms, such as the algae, this is achieved by **diffusion** and **active transport**, without the need for specialised tissues, but in the higher plants these processes alone are not sufficient and transport involves the **vascular tissues**.

VASCULAR TISSUES

The vascular tissues are made up of:

- the **xylem**, which transports water and mineral salts from the roots to the leaves
- the **phloem**, which transports the **soluble** products of photosynthesis from the leaves to all other parts of the plant

To understand how transport is achieved by these tissues, it is necessary to know about the cells which make up the tissues.

STRUCTURE OF XYLEM

Xylem is made up of four different types of cells:

- vessels
- fibres
- tracheids
- xylem parenchyma

Vessels

The vessels are **conducting** cells. A vessel is formed from a long chain of vessel segments joined end to end; the horizontal walls break down during development, leaving a long tube. **Lignin** is deposited on the walls, rendering them impermeable to gases and nutrients, so the contents die, leaving a hollow lumen ideally suited to the transport of water. The walls have a number of **pits**, areas where the lignin has not been deposited, and these areas enable lateral transport of water to occur from vessel to vessel. The deposits of lignin do provide strength and prevent the collapse of the vessels under pressure.

Tracheids

The tracheids are conducting cells and are similar to the vessels, but they are usually six-sided, with tapered ends, and retain their end walls.

Fibres

The fibres are not concerned with transport as they have very thick **lignified** walls and narrow lumens quite unsuited to that function, but more suited to support.

Xylem parenchyma

Xylem parenchyma has living contents and serves to conduct materials across the xylem tissue, as well as being a storage tissue.

TRANSPORT IN THE XYLEM

A simple experiment with a cut stem or small whole plant with its roots in a solution of dye will serve to demonstrate that the transport of water occurs in the xylem tissue. Sections of the stem will show that the xylem tissue has been stained while other tissues have not. The vessels, with their hollow lumens, offer very little resistance to the flow of water, and the properties of water itself contribute to its passage upwards in the stem. There is always a demand for water at the leaves, because of **transpiration** and **photosynthesis**, so the **adhesive** forces which exist between the water molecules and the walls of the vessels, together with the **cohesive** forces attracting the water molecules to each other, combine to maintain the columns of water in the xylem.

MOVEMENT OF WATER THROUGH THE PLANT

Water enters a plant through the **root hairs**, travels across the **cortex** and into the xylem. It then travels in the xylem up through the stem to the leaves, where most of it evaporates from the internal leaf surfaces and passes out, as water vapour, into the atmosphere.

Transpiration

Water is lost as **water vapour** from the leaves in the process known as transpiration. This process is an inevitable consequence of the opening of the stomata for gas exchange in photosynthesis. The water comes from the xylem tissue and passes into the **mesophyll** cells via three pathways:

- through the cellulose cell walls in the **apoplast** pathway
- through the cytoplasm and plasmodesmata in the **symplast** pathway
- through the **vacuolar** pathway, from vacuole to vacuole.

The first pathway depends on **mass flow** and the **cohesion** of the water molecules, the second depends on the cytoplasm of adjacent cells being interconnected and the third involves the movement of water down a **water potential gradient**. The water evaporates from the cell walls into the air spaces and then diffuses out through the stomata, from a high water potential inside the leaf to a lower one outside the leaf.

Transpiration rate

The rate at which this happens is dependent on external factors such as the **temperature**, **humidity** and **air movements**.

> **Factors affecting the rate of transpiration**

- It is possible to demonstrate that transpiration occurs by enclosing the aerial parts of a plant in a polythene bag and then testing the liquid which condenses on the inside with cobalt chloride paper.
- It is also possible to measure the rate of transpiration indirectly by measuring the water uptake of a leafy shoot using a **potometer**. This type of apparatus enables comparisons between different conditions to be made using the same shoot and between different types of plants. The movement of the air, the amount of light and also the temperature can be varied.

Other factors which affect the rate of transpiration involve the numbers and distribution of stomata, the leaf area and the thickness of the cuticle. The stomatal mechanism is referred to in Chapter 8, but it is relevant here to mention that plants living in dry habitats do show some reduction in the number of stomata, together with an increase in the thickness of the cuticle and a reduction in leaf area. These are termed **xeromorphic adaptations** and are shown particularly well by the cacti, where the leaves may be reduced to spines, and the conifers, with their needle-like leaves on which the stomata are confined to sunken grooves.

Water uptake

Water moves across the roots by the same three pathways that were mentioned when referring to the leaf:

- the **apoplast** pathway
- the **symplast** pathway
- the **vacuolar** pathway

The region of greatest uptake of water is in the **root hair zone**, where the surface area is enormously increased by the presence of root hairs from the cells of the **piliferous layer**. The root hairs come into very close contact with the soil particles; this helps to anchor the plant in the soil as well as enabling very efficient uptake of water from the soil solution. Mineral salts are taken up by active transport from the soil solution and their uptake must not be confused with that of water.

When water has entered the roots it passes across the root cortex until it reaches the **endodermis**. The cells in this innermost layer of the cortex have bands of **suberin** round them which prevent the use of the apoplast pathway, so the water and mineral ions must pass through the cytoplasm of the cells in this layer. It has been suggested that the endodermis controls the entry of water and mineral ions into the vascular tissue by secreting salts into the vascular tissues, creating a low water potential there relative to the rest of the root, so promoting the movement of water into the xylem from the cortex.

STRUCTURE OF PHLOEM

Phloem tissue consists of four types of cells:

- **sieve tubes**
- **companion cells**
- **phloem fibres**
- **phloem parenchyma**

The **phloem fibres** are more apparent in certain types of secondary phloem tissue and the **phloem parenchyma** serves as a storage tissue.

Sieve tubes

The sieve tubes are formed from **sieve elements** placed end to end (rather similar to vessels), but the end walls do not break down, but become perforated by pores, which allow substances to pass from cell to cell. These areas are known as **sieve plates** and are easily visible in both transverse and longitudinal sections of phloem tissue. When mature, the sieve tube elements do not possess a nucleus and most of the other cell organelles disintegrate, leaving a few fragments of endoplasmic reticulum and mitochondra in the thin layer of peripheral cytoplasm against the cell wall. The central part of the element is filled with **protein filaments**, which are continuous, with similar filaments in other sieve tube elements. The exact nature of the filament is not clear as it is difficult to prepare material for electron microscopy. Each sieve tube element is closely associated with at least one **companion cell** (see Fig. 11.7).

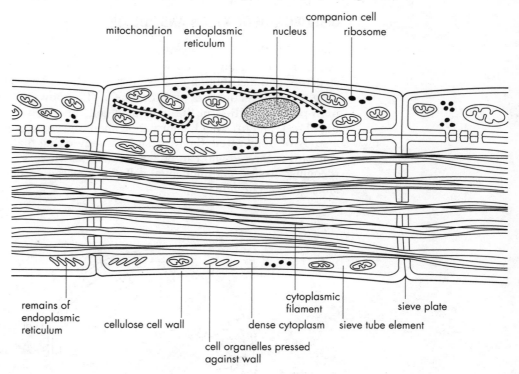

Fig. 11.7 Sieve tube and companion cell

Companion cells

The companion cells have dense cytoplasm, large centrally placed nuclei and many mitochondria; they are connected to the sieve tube elements by **plasmodesmata**.

TRANSPORT IN THE PHLOEM

The products of photosynthesis are transported in the phloem, away from the site of synthesis in the leaves to all the other parts of the plant, especially to the **apical** and **lateral meristems**, to **developing seeds** and to **perennating organs**. The products of photosynthesis must be soluble for transport and are in the form of **sucrose** and **amino acids**. It is possible to demonstrate that this transport does occur in the phloem by allowing the plant to use radioactively labelled carbon dioxide for photosynthesis and then tracing the pathway taken by the products.

Transfer cells

The sugars and amino acids enter the sieve tubes by an active process which is thought to involve transfer cells in addition to the companion cells. Transfer cells are found associated with the phloem tissue and differ from the other mesophyll cells in their structure. They move the sugars and amino acids from the mesophyll cells to the sieve tubes and, together with the companion cells, supply the necessary energy for this process.

Flow hypotheses

Several hypotheses have been put forward to account for the movement of the organic substances once in the sieve tubes, the best documented being the **mass flow** hypothesis. This suggests that there is a passive mass flow of sugars from the phloem of the leaf, where there is the highest concentration, to other areas of the plant where there are lower concentrations. This hypothesis does have its drawbacks as it suggests that all the substances move at the same rate, which they do not, and that the transport is always in one direction, which it is not.

Other hypotheses include **electro-osmosis**, **surface spreading** and **cytoplasmic streaming** along the protein filaments. Few syllabuses require candidates to have a comprehensive understanding of these alternative hypotheses, but more information can be gained from some of the standard texts and from the references given at the end of the chapter.

> Be familiar with the mass flow hypothesis

DISTRIBUTION OF XYLEM AND PHLOEM

The distribution of the **vascular tissues** differs in stems, roots and leaves, and it is a requirement of many syllabuses that candidates have some understanding of both the structure and distribution of these tissues. This is best studied using **transverse** and **longitudinal sections** of plant organs, making **low-power plans** of tissues and then **high-power studies** of the individual cells of which the tissues are composed (see Figs 11.8a) and b)). Much information about the structure of xylem can be gained from a study of **macerated tissue**: that is, woody tissue that has been treated so that the cells are separated and can be appreciated as whole structures. You must be able to recognise tissues and individual cells in prepared sections, and be able to draw and label them accurately.

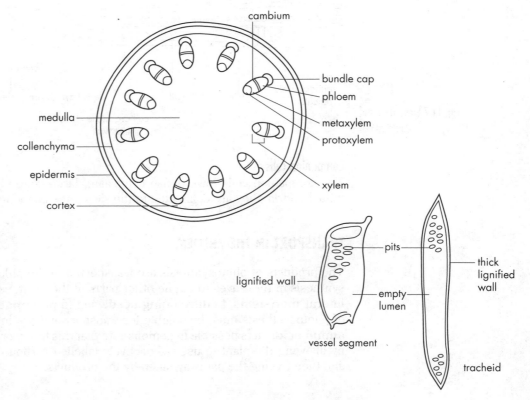

Fig. 11.8a) Low power map of tissues of young sunflower stem (transverse section)

Fig. 11.8b) High power detail of xylem cells in sunflower stem

PRACTICAL WORK

Several practical exercises have already been described in the relevant sections of this chapter, but it is worth reviewing these and emphasising those that will be most useful to the student.

Diffusion can be demonstrated in both gases and liquids, and it is easy to set up an osmometer to demonstrate osmosis. The water potential of plant tissue can be determined using different concentrations of sugar solutions; this is a favourite experiment to set in a practical test as results can be achieved in a short time. Changes in mass or linear dimensions of the tissues are used to determine the water potential.

Knowledge of the tissues and structures involved in the transport systems of both plants and animals is important, as many examination questions require a candidate to relate structure to function. Time spent looking at the histology of cardiac muscle, the structure of the heart and blood vessels and the tissues of the flowering plant will provide a firm foundation for answering questions on this section of the syllabus.

Further reading

Clegg C J and Cox G, 1978, *Anatomy and Activities of Plants*. John Murray
Bracegirdle B and Miles P H, 1971-3, *An Atlas of Plant Structure*, 2 vols. Heinemann
ABAL Unit 5, 1984, *Exchange and Transport*. Cambridge
Street H E and Opik H, 1984, *The Physiology of Flowering Plants*. Arnold

EXAMINATION QUESTIONS AND ANSWERS

QUESTIONS

Question 11.1

a) List **four** of the main functions of water in plants. *(5)*
b) When strips of potato tuber are put into distilled water they increase in length because their cells become larger.
  i) Using water potential terminology, explain how water is drawn into the plant cells. *(5)*
  ii) What ultimately limits this movement? *(2)*
c) Make a labelled diagram to show the condition of a plant cell which has been placed in a 0.8 mol dm^{-3} sucrose solution for 30 minutes. *(3)*
d) If the rate of water uptake and the rate of water loss in a herbaceous plant is measured over the same periods, results such as those shown in the graph may be obtained (see Fig. 11.9, p. 144).
Explain what has happened to the water content of the plant over the period shown. *(4)*
e) Describe the likely appearance of the plant at 16.00 hours. *(1)*

(Total 20 marks)

Question 11.2

a) With the aid of labelled diagrams, describe the structures involved in the translocation of organic solutes between different parts in an Angiosperm plant. *(10)*
b) Describe the mechanisms that have been suggested to account for the movement of solutes during translocation. Which do you consider the most likely explanation or explanations? *(14)*
c) Explain **one** experiment which has been performed that has enhanced our knowledge of translocation in flowering plants. *(6)*

(Total 30 marks)
(ODLE)

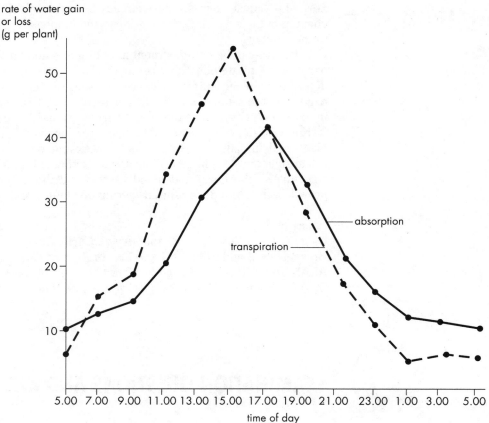

Fig. 11.9

Question 11.3
Diagrams A, B and C below show cross-sections of three different types of blood vessel.

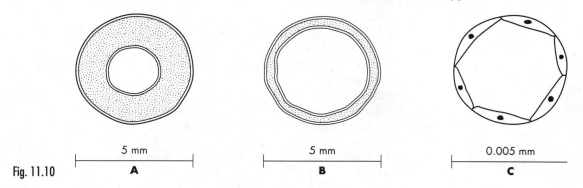

Fig. 11.10

They are not drawn to the same scale.
a) Identify blood vessels A, B and C. *(3)*
b) State **two** ways in which vessel A is adapted for its functions. *(2)*

(ULEAC)

Question 11.4
a) Draw a large, labelled diagram to show the structure of an artery as seen in cross-section. *(6)*
b) Give **four** differences in structure between arteries and veins. Write your answers in the table below. *(4)*

Arteries	Veins

(Total 10 marks)
(ULEAC AS-level)

Question 11.5

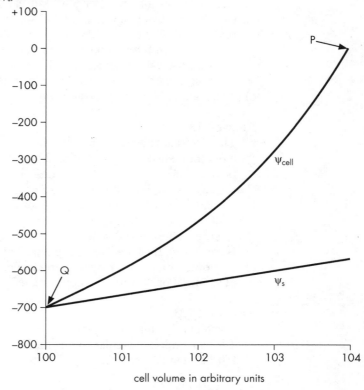

Fig. 11.11

The graph shows the relationship between the volume, water potential (ψ_{cell}) and solute potential (osmotic potential) (ψ_s) of a plant cell immersed in a series of sucrose solutions of increasing concentration. In each solution the cell was allowed to reach equilibrium with the bathing solution, so that water was being rather lost not gained, before the measurements were made.

a) What terms are used to describe the condition of the cell at (i) P and (ii) Q? (2)
b) What is the volume of the cell (in arbitrary units) when in equilibrium with a sucrose solution of solute potential – 400kPa? (1)
c) Calculate the pressure potential (turgor pressure) (ψ_p) of the cell when its volume is 103 units. Show your working. (2)
d) Suggest **two** ways in which reversible changes in cell volume may be important in flowering plants. (2)

(Total 7 marks)
(ULEAC)

OUTLINE ANSWERS

Answer 11.1

a) 1 transport of ions in solution
   2 transport of organic solutes such as sucrose
   3 maintaining turgidity of guard cells for stomatal opening
   4 support from turgid parenchyma tissue in young tissues
b) i) The cell sap in plant cells has solutes in it which makes its water potential lower, or more negative, than the distilled water. Water is taken up by the tissue, by osmosis, as water will move from a high water potential to a lower one. The entry of water into the cells of the tissue will increase the volume of each cell, thus increasing the overall length of the potato strips.
   ii) This movement of water will be limited by the cellulose cell wall, which will exert a pressure against the protoplast of the cell. No more water can be taken up when the water potential of the cell is equal to the solute potential together with the pressure exerted by the wall.

c) The concentration of the sucrose solution is high, so this drawing should show a plasmolysed cell with the cell contents rounded up in the middle, having little or no contact with the cell wall.
d) Gradually during the morning and early afternoon, the rate of transpiration has exceeded the rate of water uptake, so the plant will be losing more water than it can replace. At about 17.00 hours the two rates are the same. This probably coincides with a drop in temperature or light intensity, so the rate of transpiration will begin to drop; the rate of absorption is always more than the transpiration rate, so the water content decreases and then increases.
e) The plant is likely to be wilted, with drooping leaves, at 16.00 hours.

Answer 11.2
a) The diagrams needed would be of phloem tissue, a sieve tube and a companion cell; a diagram showing the relationship of the vascular tissue with transfer cells could be helpful as well. A written description of these cells and their structures is required as well as the diagrams.
b) Explanations of the mass flow hypothesis and of cytoplasmic streaming, together with more up-to-date theories such as the electro-osmosis and transcellular strand theories, should be discussed.
Note: These theories are not required in all syllabuses – check yours.
c) Reference to an experiment using radioactive tracers, or aphids, or to the removal of a ring of bark would be appropriate here.

Answer 11.3
a) A artery; B vein; C capillary.
b) 1 It has a thick muscular tunica media.
2 It has a small lumen.

Answer 11.4
a) Marks would be awarded for the size and quality of the drawing. The correct thickness of the wall must be shown. Labels should include: endothelium/tunica intima; muscle layer/tunica media; fibrous layer/tunica externa; elastic fibres; lumen.
b) Many of the points appear in the answer given in Question 11.5 below. The points made must be structural and not functional There must be a relevant comment in both columns for the mark to be given.

Answer 11.5
a) i) P turgid
ii) Q flaccid or plasmolysed
b) 102.4
c) Give the equation: $\psi_{cell} = \psi_P + \psi_s$
then substitute values.
$\psi_P = -280\text{kPa} - -600\text{kPa}$
$= 320\text{kPa}$
d) Opening and closing of stomata; nastic movements in petioles; wilting.

QUESTION, STUDENT ANSWER AND EXAMINER COMMENTS

Question 11.5
Distinguish between each member of the following pairs:
a) artery and vein

> 66 Correct distinguishing point 99
> 66 Unnecessary to put in that both have these layers, and it is only the muscular layer of the artery that is thicker, the fibrous layers in both are of similar thickness 99
> 66 Correct answer 99

```
An artery carries blood from the heart; a vein carries
blood towards the heart. Both have a muscular layer and a
fibrous layer in the wall. However, the artery has both
layers much thicker than the vein. The artery has a narrow
lumen, the vein has a wide lumen. The vein has valves
across the lumen to prevent backflow, the artery does not
have valves.
```

(4)

b) plasma and intercellular fluid

> Plasma is the liquid base of the blood. It contains dissolved food materials, waste products, hormones and plasma proteins. Intercellular fluid is formed from blood by ultrafiltration and has the same constituents as plasma less the plasma proteins.

(2)

There is enough information here for 2 marks. Student could have made reference to the fact that plasma is contained within the blood vessels, intercellular fluid bathes the cells of the tissues, as an alternative answer.

c) antigen and antibody

Good answer

> An antigen is a macromolecule foreign to the body, which initiates an immune response. An antibody is a protein (corresponding to an antigen) which combines with the antigen and destroys the foreign particle.

(2)

d) lymphocyte and neutrophil granulocyte (= polymorphonuclear leucocyte)

This is a very good answer, giving differences in both structure and function

> A lymphocyte has a spherical nucleus and clear cytoplasm. A neutrophil has a lobed nucleus and granular cytoplasm. Neutrophils are phagocytic and capable of amoeboid movement. Lymphocytes produce antibodies.

(2)
(Total 10 marks)
(ODLE)

CHAPTER 11 REVIEW SHEET

REVIEW SHEET

1 What do you understand by **diffusion**?
_____
_____

2 Identify three factors affecting the **rate of diffusion**.
  a) _____
  b) _____
  c) _____

3 Draw a diagram and use it to demonstrate the existence of osmosis.

4 Define **active transport**. Briefly explain the process by which it occurs, using a diagram by way of illustration.

5 Complete the labels on the diagram below; then use your diagram to explain **blood flow**.

	RA (right atrium)	LA (left atrium)	
	RV (right ventricle)	LV (left ventricle)	

_____ blood side   _____ blood side

6 Complete the labels of the diagram below; then use your diagram to explain the mechanisms involved in the **contraction of the heart**.

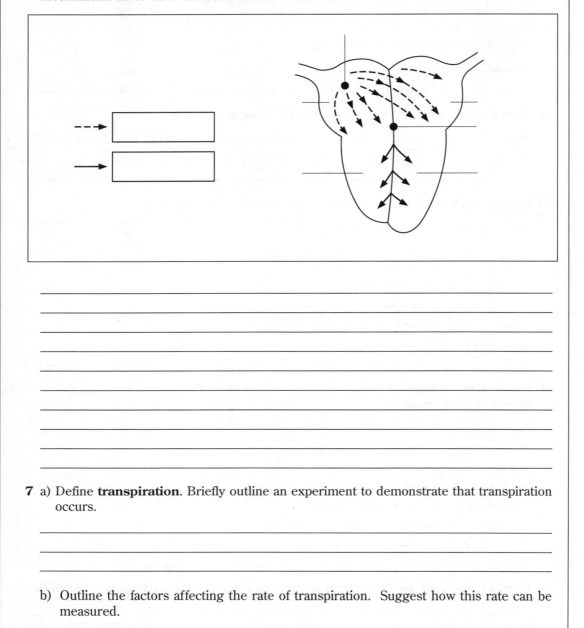

7 a) Define **transpiration**. Briefly outline an experiment to demonstrate that transpiration occurs.

b) Outline the factors affecting the rate of transpiration. Suggest how this rate can be measured.

8 Describe each of the four types of cells which make up the **xylem**.

a)

b)

c)

d)

9 Briefly describe an experiment to demonstrate that **transport of water occurs in the xylem tissues**.

_____
_____
_____
_____

10 Complete the labels of the diagram below.

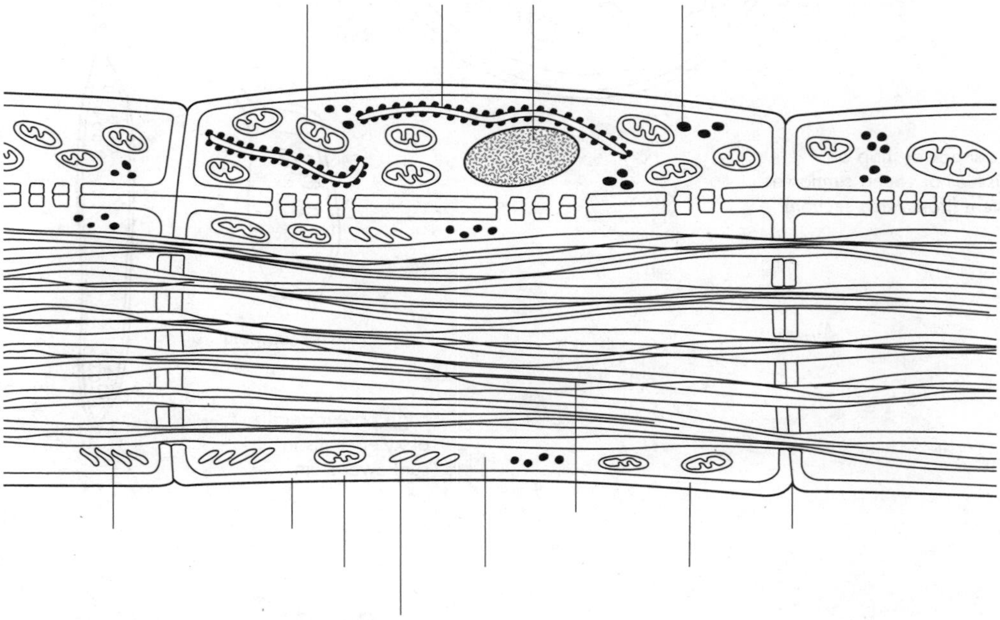

Sieve tube and companion cell

11 Briefly explain how **transport occurs in the phloem**.

_____
_____
_____
_____
_____
_____
_____

12 Complete the labels in the diagram below.

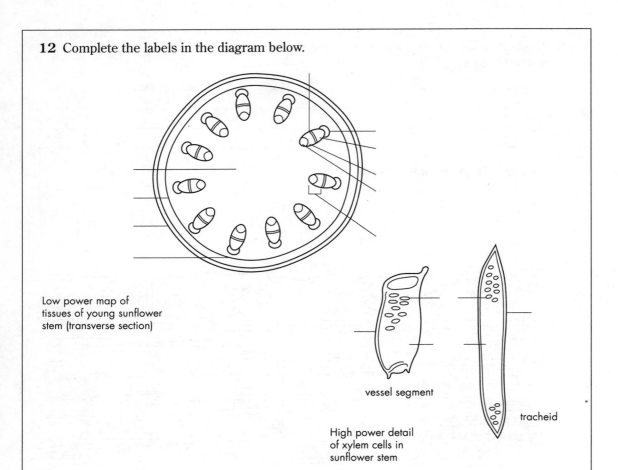

Low power map of tissues of young sunflower stem (transverse section)

vessel segment

High power detail of xylem cells in sunflower stem

tracheid

CHAPTER 12

REPRODUCTION AND DEVELOPMENT IN MAMMALS

MALE REPRODUCTORY SYSTEM

FEMALE REPRODUCTORY SYSTEM

GAMETOGENESIS

FEMALE SEXUAL CYCLES

TRANSFER OF GAMETES

FERTILISATION

ARTIFICIAL CONTROL OF FERTILISATION

EARLY EMBRYONIC DEVELOPMENT

EMBRYONIC MEMBRANES

FETAL CIRCULATION

BIRTH, LACTATION AND PARENTAL CARE

PRACTICAL WORK

GETTING STARTED

Reproduction in mammals is included in all the A-level syllabuses. It is a valuable study, as mammalian reproduction includes not only the production of gametes and fertilisation but internal development and the most highly developed system of post-natal care. Some syllabuses do not include any study of embryological development other than implantation and the formation of the placenta. Questions on reproduction may include parts concerned with mitosis or meiosis, so you should refer to those sections in Chapter 15. In your coursework you may have to dissect a mammalian reproductive system and/or carry out a histological study of the ovary and testis. It would be useful, therefore, to refer to or revise your practical work in these topics alongside this chapter. As most syllabuses use the human as the type mammal, this chapter will refer largely to human reproduction.

Parental care is an important aspect of mammalian reproduction. This is not only important for the survival of the offspring but also in many species forms the basis for learning about the social organisation of the group. You should refer back to social behaviour in Chapter 10 when using this section of the chapter. Another important aspect of human reproduction is the ability to control fertility. This includes not only preventing unwanted births but also helping infertile couples to reproduce.

CHAPTER 12 REPRODUCTION AND DEVELOPMENT IN MAMMALS

ESSENTIAL PRINCIPLES

MALE REPRODUCTIVE SYSTEM

STRUCTURE AND FUNCTION

The male system consists of a pair of **testes**, which produce **spermatozoa**; the **penis**, which is an intromittent organ; genital ducts connecting the two; and various accessory glands which provide constituents for the semen.

Each testis consists of about 1,000 **seminiferous tubules** supported by connective tissue. The latter contains **interstitial (Leydig) cells**. The whole structure is enclosed in a capsule. The seminiferous tubules produce the **spermatozoa** whilst the Leydig cells produce the male hormone, **testosterone**. The testes are housed in the scrotal sac below the abdomen and are maintained at a temperature about 2°C below the core body temperature. In order to maintain the ideal temperature for sperm production, the testes may be moved nearer to or further away from the abdomen by means of the **dartos muscle**. The structure of a testis is shown diagrammatically in Fig. 12.1.

> *Leydig cells are stimulated by luteinising hormone*

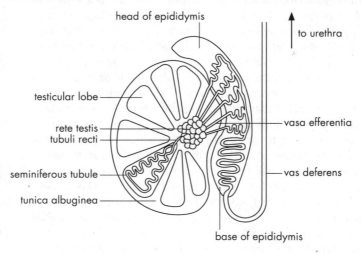

Fig. 12.1 Structure of a testis

When sperm have been produced they collect in the **vasa efferentia** and then pass to the head of the **epididymis**, where they mature. They then pass along the coiled tube to the base of the epididymis, where they are stored for a short time before passing via the **vas deferens** to the uretha during ejaculation.

FEMALE REPRODUCTIVE SYSTEM

STRUCTURE AND FUNCTION

There are two ovaries, each about the size of an almond. Each consists of a medulla and cortex and is surrounded by a connective tissue sheath. The ova are produced in the **germinal epithelium** which is the outer layer of the cortex. Once formed, the ova develop in follicles in the cortex. The medulla is made up of a mixture of connective tissue, blood vessels and mature follicles forming the **stroma**. Mature follicles migrate back to the surface for the ova to be shed. The structure of an ovary is shown in Fig. 12.2.

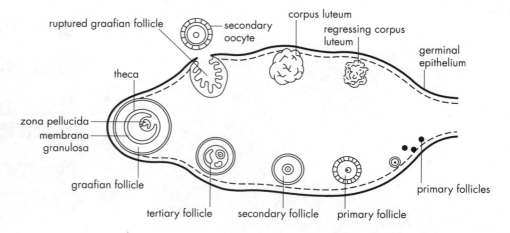

Fig. 12.2 Structure of an ovary

Ova when shed are collected by the **fimbriae**, the fringe-like end of the **fallopian tube** (oviduct), which conveys them to the uterus. The passage of the ova is helped by the ciliated epithelium of the tube and by the peristaltic contractions of the muscle in its wall.

The uterus is a muscular organ about 7.5 cm long and 5.0 cm wide; it has a thin serous coat, a thick muscular wall (**myometrium**) and a glandular inner layer (**endometrium**). The entrance to the uterus (**cervix**) opens into the **vagina**, which in turn opens to the outside in the **vulva**. This is formed by two folds of tissue, the **labia**, which also enclose the urethral opening and the clitoris. In the walls of the vulva are the **vestibular glands** which produce mucus when the female is sexually aroused, and this helps to lubricate the penis during intercourse.

GAMETOGENESIS

> Be sure you know at which stage meiosis I and meiosis II take place

The cells of the germinal epithelium of both the testis and the ovary undergo a sequence of mitotic and meiotic divisions to form gametes. First there is a multiplication stage which involves repeated mitotic divisions to produce **spermatogonia** and **oogonia**. In the male, this process is continuous from puberty throughout life, whilst in the female all the oogonia are produced before birth in the fetal ovary. Once formed, the spermatogonia and oogonia grow to full size and then undergo maturation which involves meiotic division and then differentiation into the mature gametes. For details of meiosis, refer to Chapter 15. Fig. 12.3 shows the sequence of events in gametogenesis.

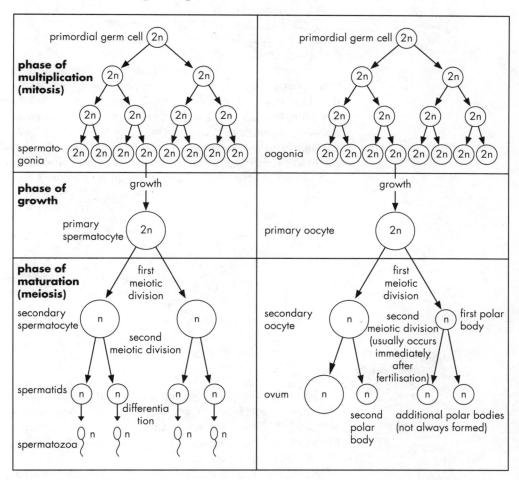

Fig. 12.3 Gametogenesis

FORMATION OF SPERMATOZOA

Spermatogonia are produced in the germinal epithelium of the seminiferous tubule and mature in their deeper layers, nourished and protected by the **Sertoli cells**. They form **primary spermatocytes**, which then form haploid **secondary spermatocytes** after the first meiotic division. After the second meiotic division they form **spermatids**, which become embedded in the folds of the Sertoli cells and develop into **spermatozoa**. The Sertoli cells provide nutrition for the developing spermatozoa and confer immunity to antigenic reaction in the male and female tracts.

FORMATION OF OVA

A similar sequence of events takes place in the formation of ova, but the detail and time sequence are different. The oogonia, which are formed before birth, undergo mitosis to form **primary oocytes**, which then mature just before ovulation. Each primary oocyte is surrounded by a single layer of cells, the **membrana granulosa**, forming the **primordial follicle**. About 2 million of these are formed in the fetal ovary but only about 450 will later develop into **secondary oocytes**. Before ovulation, a primary oocyte undergoes the first meiotic division to form the haploid secondary oocyte and the first polar body. The secondary oocyte begins the second meiotic division, but this is arrested at metaphase unless fertilisation takes place. On fertilisation, this division is completed to form a large **ovum** and the second polar body. Once this division has taken place the nucleus of the ovum fuses with that of the sperm to form a **zygote**, which will develop into an embryo.

FEMALE SEXUAL CYCLES

In all mammals female sexual activity is cyclical. A cycle in the ovary leads to the production of mature ova and a uterine cycle prepares the uterus to receive fertilised eggs. Details differ from one mammal to another and regular complete preparation of the uterine lining, followed by menstruation if the egg is not fertilised occurs only in humans. Both the ovarian and uterine cycles are controlled by ovarian hormones, which in their turn are controlled by pituitary gonadotrophins.

At the beginning of a cycle in the ovary, primary follicles begin to grow under the influence of **follicle stimulating hormone (FSH)**. One of them develops into a mature **Graafian follicle**, which contains a secondary oocyte. At ovulation the secondary oocyte detaches from the follicle wall and passes into the fallopian tube, protected by two layers of follicle cells, an inner **zona pellucida** and an outer **corona radiata**. The development of the follicles is influenced by **luteinising hormone (LH)**. Once ovulation has taken place the level of LH falls and **prolactin** from the pituitary brings about the conversion of the ruptured follicle to a **corpus luteum**, which releases **progesterone**. The presence of progesterone, with smaller amounts of **oestrogen**, maintains the receptive state of the uterine wall and by negative feedback prevents the production of FSH and LH and consequently the release of more ova.

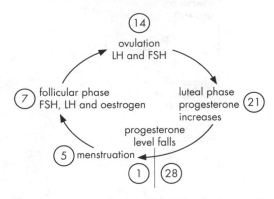

Fig. 12.4 Stages in the human menstrual cycle

Fig. 12.4 shows the stages in the human menstrual cycle. Graphs showing the changes in levels of these hormones will be found in standard biology texts. If fertilisation does not take place, the corpus luteum regresses to form a **corpus albicantus**. The levels of oestrogen and progesterone fall, FSH is again released and the cycle begins again.

Whilst the ovarian cycle is taking place, there is also a cycle of change in the uterine wall. During the follicular phase of the oestrus cycle, under the influence of oestrogen, the endometrium proliferates and thickens. Once the corpus luteum is formed, the progesterone it secretes maintains the endometrium in a receptive state and stimulates the mucus glands in it to secrete. If fertilisation does not take place, the level of progesterone falls and the thickened epithelium of the endometrium is shed, producing the menstrual flow.

CHAPTER 12 ESSENTIAL PRINCIPLES

TRANSFER OF GAMETES

Mammals have internal fertilisation and to facilitate this have developed an intromittent organ, the penis. On sexual arousal, this is erected by the increase of blood pressure in the erectile tissue. **Parasympathetic** stimulation narrows the venules and dilates the arterioles to bring this about. The penis is inserted into the female's vagina and the stimulation by friction produced by rhythmic movements brings about ejaculation of semen into the vagina. Ejaculation is a **sympathetic nervous system** reaction. The internal sphincter of the bladder is closed and smooth muscle in the male system and accessory glands contracts to discharge the sperm and seminal fluids. The semen consists of a mixture of fluids from the various accessory glands and contains about 10^8 sperms cm^{-3}. Seminal fluid is alkaline in nature and raises the pH of the female tract to about pH 6.5, which is the optimum for sperm motility.

FERTILISATION

Sperm are deposited near to the cervix and quickly pass to the top of the fallopian tube by contractions of the uterus and fallopian tube. The contractions are thought to be due to the action of **oxytocin** and **prostaglandins**. Sperm are highly fertile for 12 to 24 hours after release into the female tract but can fertilise an ovum only after a process called **capacitation**, which takes several hours. It involves changes in the membrane covering the **acrosome**, a thin cap over the nucleus. The structure of a sperm is shown in Fig. 12.5.

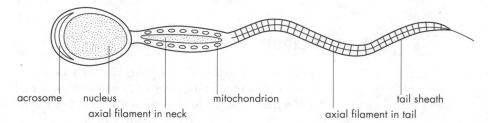

Fig. 12.5 Stucture of a human sperm

When sperm reach an oocyte the acrosome membranes rupture and two enzymes, **hyaluronidase** and **protease**, are released and digest away the layers of cells surrounding the oocyte. Further changes in the sperm head allow it to enter the oocyte. This entry stimulates reactions in the zona pellucida of the oocyte which brings about the formation of the **fertilisation membrane**, preventing the entry of further sperm. Entry of the sperm also stimulates the completion of the second meiotic division of the oocyte nucleus. The nuclei of the ovum and sperm (known as **pronuclei**) are drawn together and fuse to form a diploid nucleus. The fertilised ovum is known as the **zygote** and it begins to divide by mitosis to form the embryo.

ARTIFICIAL CONTROL OF FERTILISATION

There are two aspects to this: the prevention of births by contraception and abortion; and giving help to couples who are unable to have children.

CONTRACEPTION

Without any birth control, the theoretical number of pregnancies a woman would have between puberty and menopause is about twelve. Various methods are adopted to limit the number of pregnancies. In some cultures abstinence and late marriage are used, in others lactation is extended to reduce the likelihood of pregnancy. Some religious groups do not agree with artificial methods of birth control; they make use of the rhythm method or 'safe periods'. The idea is to limit intercourse to the portion of the month when conception is least likely. If periods are regular in length, the 'unsafe' days can be calculated fairly accurately but this is not so if the woman has periods of varying length. Daily records of temperature may help, as at ovulation there is a slight decrease, 0.5°C, in early morning temperature, with a slight rise, 0.8°C, the day after.

Artificial methods
Artificial methods range from coitus interruptus, where the penis is withdrawn before ejaculation, the least effective method, to the oral contraceptive, which, if taken regularly, is almost 100% effective. In between are the barrier methods using a male sheath or a female diaphragm. When used with a spermicidal cream or jelly, these methods are about

90% effective: that is, one would expect 10 pregnancies per 100 women per year. Of similar effectiveness is the intra-uterine device, which probably prevents implantation.

Sterilisation
For couples who have completed their desired family, sterilisation may be used. In the male this is a very simple operation which involves cutting the vasa deferentia (**vasectomy**). In the female the operation is more complex, involving the ligation or cauterisation of the fallopian tubes. Either of these is 100% effective but must be assumed to be a permanent change.

ABORTION

This is the removal of the developing embryo from the uterus, usually by suction or curettage. In many countries with well-developed medical facilities, abortion is a very widely used method of birth control. In Britain the number of abortions has increased greatly over the past 20 years and there are now in excess of 100,000 abortions annually. Some abortions follow **amniocentesis** if this technique has shown an embryo to be malformed. This is in fact a method of avoiding the birth of babies with congenital defects and is not a method of birth control.

INFERTILITY

About 1 in 10 couples are infertile. This may be due to a low sperm count or blockage of the vas deferens in the male or to blockage of the fallopian tubes or failure to ovulate in the female. Failure to ovulate may be due to a thickened ovarian capsule but is more likely to be the result of a hormone imbalance.

Treatment
Little can be done to improve male fertility. If in women the problem is failure to ovulate, it may be cured by giving FSH or other gonadotrophins to stimulate maturation of follicles. Dosages have to be carefully regulated to lower the likelihood of multiple births. In the case of blocked tubes, it is possible to remove ova from the ovaries by suction, fertilise them *in vitro* and then implant the fertilised ova into the woman's uterus. This technique is still in the developmental stage even though many 'test-tube' babies have now been born. Few of the *in vitro* fertilisation clinics have more than a 20% success rate.

EARLY EMBRYONIC DEVELOPMENT

Division
As soon as an ovum is fertilised it begins to divide to form a ball of cells, the **morula**. As cleavage continues, a fluid-filled cavity develops inside the bundle of cells. This stage is the **blastula**, the cells forming it are **blastomeres** and the cavity is the **blastocoel**. The outer layer is the **trophoblast**. At one point it thickens to form an inner cell mass. The structure of a blastocyst is shown in Fig. 12.6. This process occurs during the ovum's passage down the fallopian tube.

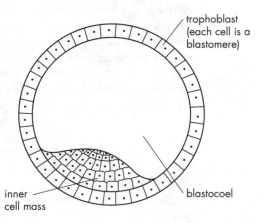

Fig. 12.6 Structure of a blastocyst

Implantation
After reaching the uterus, the **zona pellucida** disappears and the blastocyst embeds in the endometrium, usually 6–9 days after fertilisation. This is **implantation**.

Differentiation

The blastocyst continues to swell by absorbing uterine fluid. At the same time the inner cell mass flattens to form the **embryonic disc**, consisting of several layers of cells. Cells from the lower layer spread out to form a complete layer inside the trophoblast enclosing the fluid. The trophoblast over the embryonic disc breaks down so that the disc comes to lie on the surface. From the outer surface of the disc, cells migrate upwards to form the mesoderm, which divides into somatic and lateral plate regions. A split appears in the lateral plate, dividing it into **somatic** (next to the ectoderm) and **splanchnic** (next to the endoderm) mesoderm. The fluid-filled space between the two layers forms the coelom. This mesoderm extends only about halfway down the blastocyst. Other cells migrate inwards from the upper surface to form the notochord, whilst on the surface the **neural plate** forms. This folds inward to form the neural tube, the primitive central nervous system.

EMBRYONIC MEMBRANES

Four embryonic membranes develop:

- the **yolk sac**
- the **amnion**
- the **chorion**
- the **allantois**

The yolk sac

The yolk sac consists of endoderm overlain by mesoderm in the embryonic part and by ectoderm (trophoblast) in the lower part. The part immediately below the embryo gives rise to the fore-gut and hind-gut, whilst the lower part develops **trophoblastic villi**, which grow into and absorb nutriment from the uterine wall.

The amnion and chorion

> Do not confuse these two membranes

The amnion and chorion arise together as a fold of the **somatopleur** (mesoderm and trophoblast) around the edge of the embryonic disc. The amnion consists of endoderm with mesoderm outside, whilst the chorion has the layers reversed, mesoderm on the inside. The cavity between them is the **extra-embryonic coelom**. As soon as the chorion is complete, it forms **chorionic villi**, which burrow into the uterine wall.

The allantois

The **allantois** grows out from the developing hind-gut of the embryo and consists of an outer layer of mesoderm and an inner layer of endoderm. It is concerned with respiration and excretion prior to the formation of the placenta. Around the tail of the embryo it fuses with the chorion to form the **allanto-chorion**. These structures are shown in Fig. 12.7.

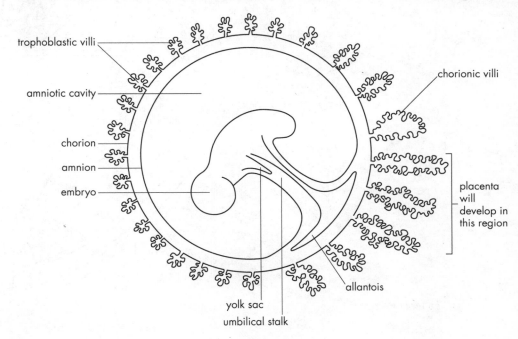

Fig. 12.7 Embryonic membranes

THE PLACENTA

> Students often confuse the functions of the amnion and amniotic fluid with those of the placenta — make sure you are clear about their functions

The placenta is formed in the area where the allanto-chorion attaches to the wall of the uterus. Villi from the embryonic structure interlock with villi from the uterine wall. Initially, there are six layers of tissue between the maternal and fetal blood but, as the placenta develops, some of these disappear so that in the mature placenta of a human, the fetal capillaries are separated from the maternal blood only by the trophoblast and embryonic mesoderm. This provides a minimal barrier to the diffusion and active transport of materials across the placenta. Water, amino acids, glucose, lipids, mineral salts, simple proteins, vitamins, hormones, antibodies and oxygen pass from mother to fetus. Water, carbon dioxide, nitrogenous waste and hormones pass in the other direction. Unfortunately, the easy passage of materials also allows the passage of potentially harmful bacteria, viruses, toxins and drugs into the fetus.

The other main function of the placenta is to protect the fetal circulation from the high maternal blood pressure.

FETAL CIRCULATION

> This can be difficult to understand. Study this section carefully

During gestation

During gestation all nutrition and gas exchange take place via the placenta, the fetal digestive tract and lungs being non-functional. Most of the oxygenated blood entering the fetus via the umbilical vein is shunted to the inferior vena cava via the **ductus venosus**, bypassing the fetal liver. The rest of the blood from the umbilical vein flows straight to the liver and then to the ductus venosus. Blood entering the right atrium is therefore a mixture of oxygenated and deoxygenated blood. From the right atrium, most of the blood passes through the **foramen ovale**, an opening in the atrial septum, into the left atrium. Blood going from the right atrium to the right ventricle does not pass to the lungs but is shunted via the **ductus arteriosus** into the aorta. In the fetus blood pressure is greatest in the pulmonary artery and this determines the direction of blood flow through the fetus and placenta. This circulation is shown in Fig. 12.8.

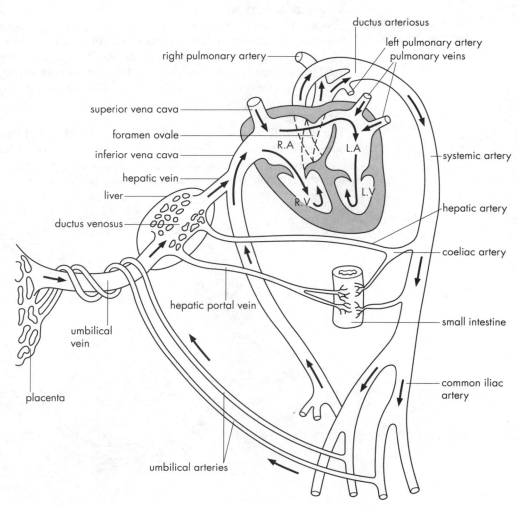

Fig. 12.8 Fetal circulation

At birth

At birth the lungs suddenly inflate, reducing the resistance to blood flow, so that blood passes through the lungs instead of through the ductus arterosus, so reducing the pressure in the pulmonary artery. Tying the umbilical cord stops the flow of the baby's blood through the placenta, increases the blood volume in the baby's body and so reduces the systemic blood pressure. The foramen ovale closes, separating the left and right atria. The ductus venosus also closes, increasing blood flow through the liver, thus completing the change from the fetal to the post-natal circulation.

BIRTH, LACTATION AND PARENTAL CARE

During gestation the levels of progesterone and oestrogen increase until immediately before birth, when the progesterone level falls. It is thought that some as yet unidentified stimuli cause the **fetal pituitary** to secrete ACTH, which in turn causes the **fetal adrenal gland** to secrete corticosteroids. These cross the placental barrier, causing the decrease in progesterone secretion, an increase in the production of **prostaglandins** and the release of **oxytocin** from the maternal pituitary. The oxytocin causes contraction of the smooth muscle in the uterus and the prostaglandins increase the strength of the contractions. These contractions are wavelike, each wave being caused by a release of oxytocin. The waves pass from the top to the bottom of the uterus. They are accompanied by the dilation of the cervix and the rupture of the embryonic membranes, which release the amniotic fluid from the cervix. Shortly after birth, the uterus contracts vigorously, expelling the placenta and sealing the blood vessels which supplied the placenta with maternal blood. This process minimises maternal blood loss after the birth.

LACTATION

> Be clear about the functions of prolactin and oxytocin

Mammary glands consist of glandular epithelial cells lining **alveoli**, which form 15 lobules in the breast. Each alveolus is surrounded by a layer of contractile tissue and the lobules are supported by a layer of connective tissue. Milk is produced by the glandular cells and is stored in sinuses which are connected to openings in the nipple. During pregnancy the breasts increase in size under the influence of **prolactin** but the presence of progesterone inhibits the production of milk. When the progesterone level falls at birth, prolactin stimulates the secretion of milk; the presence of oestrogen is also necessary for milk production. Suckling stimulates a nervous response which, via the hypothalamus, causes the release of oxytocin. This in turn causes the contraction of the **myoepithelium** round the alveoli, forcing the milk from the sinuses out of the nipple. For the first few days after the birth, the breasts produce **colostrum** not milk. This is rich in protein and antibodies.

PARENTAL CARE

As well as being provided with milk, mammalian babies need other forms of care. Babies have a large surface area to volume ratio and so lose heat quickly. Therefore human babies, which are born naked, must be kept warm. Babies and juvenile animals are easy prey for predators so most mammal species have some system of protection of their young. In humans parental care also includes teaching the young child how to look after itself and how to live in society. Other mammals give protection when their offspring are very young but learning is usually just by observation of the parents. Mimicry of the parents and play amongst siblings are both important in the learning of survival techniques.

PRACTICAL WORK

For the structure of the male and female systems you need to dissect a small mammal or to study anatomical models. Microscope work is necessary to study the structure of an ovary and a testis. Sections should be observed under high power to see the detailed biology of the two organs. If your syllabus includes early embryology, models will be useful for you to understand the three-dimensional structure of an embryo and its membranes. As an adjunct to mammalian embryology, it would also be useful to look at sections of a three-day chick embryo to see the relationship of the allantois, yolk sac, amnion and chorion.

If your syllabus includes parental care, visits to an ante-natal clinic and a child welfare clinic would be useful, as would talks with a health visitor or paediatrician.

Further reading
Cohen J and Massey B, 1984, *Animal Reproduction*. Arnold
Gadd P, 1983, *Individuals and Populations*. CUP
Green N P O, Stout G W and Taylor D J, 1985, *Biological Science Vol.2*, Ch.20, Reproduction, Ch.21, Growth and development. CUP
Newth D R, 1970, *Animal Growth and Development*. Arnold

EXAMINATION QUESTIONS AND ANSWERS

QUESTIONS

Question 12.1
The diagram below shows the relationship between:
  i) the pituitary gonadotrophins
  ii) the ovarian steroids
  iii) follicle development
  iv) the thickness of the endometrium during the human oestrus cycle.

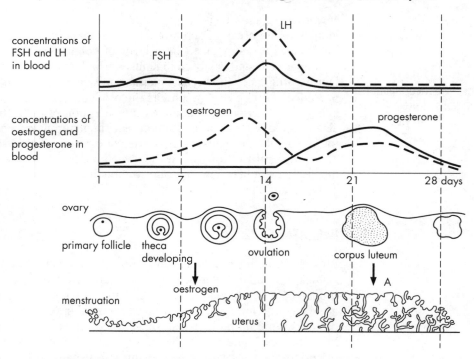

Fig. 12.9

a) From which cells is the hormone oestrogen secreted and what evidence is there for this in the diagram? (2)
b) Give **one** effect of the hormone oestrogen on the pituitary gland. Explain your answer with reference to the diagram. (2)
c) What is represented by the arrow A on the diagram? (1)
d) What evidence is there from the diagram that the hormone progesterone is involved in a negative feedback mechanism? (2)
e) How is a knowledge of the effects shown in the diagram involved in the formulation of the contraceptive pill? (2)

(Total 9 marks)
(ODLE)

CHAPTER 12 EXAMINATION QUESTIONS

Question 12.2
The diagram below is of mammalian ovary.

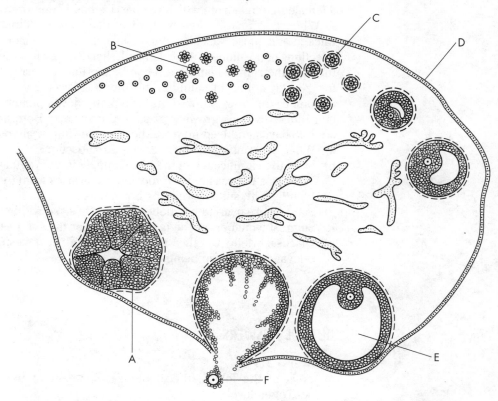

Fig. 12.10

a) Which of the following structures does the letter A indicate? Underline the correct answer.
   i) Interstitial cells
   ii) Corpus luteum
   iii) Primary follicle
   iv) Secondary oocyte
   v) Germinal epithelium   (1)
b) Which of the following is the correct sequence in the development of the labelled structures? Underline the correct answer.
   i) AFEDCB
   ii) EFBCDA
   iii) CDBAFE
   iv) DBCEFA
   v) DABCEF   (1)
c) For each of the following, state whether it is haploid or diploid.
   i) Secondary oocyte
   ii) Primary oocyte
   iii) Germinal epithelium
   iv) Oogonium   (4)
d) i) Name **one** hormone produced by the ovary.
   ii) What is the main role of this hormone?   (2)
e) State **three** general features of animal hormones.   (3)
f) What part does the placenta play in the
   i) nutrition of the embryo?   (3)
   ii) protection of the embryo?   (3)

(Total 17 marks)
(ULEAC)

Question 12.3
a) Name two hormones involved in birth. For each state **one** effect.   (4)
b) What is the role of hormones in lactation?   (2)

(Total 6 marks)
(ULEAC)

CHAPTER 12 REPRODUCTION AND DEVELOPMENT IN MAMMALS

Question 12.4
The effect of ovarian hormones on the endometrium of the uterus has been studied in a female mammal from which the ovaries had been removed.

When hormone X alone was given daily, the endometrium became progressively thicker and more vascular, but development was incomplete. After 28 days, if the injections of hormone X were stopped, then 2 days after the last injection bleeding from the endometrium began. In a second experiment, hormone X was again administered for 28 days, but in addition hormone Y was given between days 15–28. The endometrium now showed its complete development, with coiled arteries with characteristic loops growing close to the surface. Both hormones were discontinued after 28 days and bleeding from the endometrium again occurred after 2 days.

a) What is meant by the term endometrial tissue?
b) State the significance of the blood supply in the regulation of the endometrium.
c) What was the reason for removing the ovaries from the experimental animal?
d) Identify with suitable reasons hormones X and Y.
e) State where ovarian hormones are synthesised within the ovary.
f) Suggest which organelle is responsible for the synthesis of ovarian hormones.
g) Discuss briefly whether the results from the investigation outlined above have relevance to human beings.

(Total 14 marks)
(UCLES)

OUTLINE ANSWERS

Answer 12.1
a) Cells of the follicle/theca; the development of the follicle mirrors the rise in oestrogen level
b) Inhibits FSH production; fall in FSH with rise in oestrogen
   OR
   Promotes LH production; positive correlation between oestrogen and LH levels
c) Maintenance of the endometrial lining
d) While there is a high level of progesterone the pituitary does not produce FSH or LH; when the progesterone level falls the pituitary starts to secrete gonadotrophins again
e) Pill contains progesterone, which inhibits FSH; prevents development of ova

Answer 12.2
a) Corpus luteum
b) iv) DBCEFA
c) i) haploid   ii) diploid   iii) diploid   iv) diploid
d) Oestrogen; female secondary sexual characteristics/oestrous cycle
   OR
   Progesterone; inhibition of ovulation/maintenance of uterine lining
e) Produced in endocrine glands; bring about an effect on a target organ; have a specific regulatory effect
f) i) Brings fetal blood into close proximity with maternal blood; provides a large surface area for exchange; minimal barrier for rapid exchange of nutrients into fetal blood
   ii) Protects the fetus from high maternal blood pressure; allows the passage of antibodies from the mother to give passive immunity; produces progesterone to maintain the pregnancy

Answer 12.3
a) i) Oxytocin; causes contraction of smooth muscle of myometrium
   ii) Prostaglandins; increase power of contractions
b) Prolactin stimulates the secretion of milk; oxytocin causes the contraction of muscles to force the milk through the nipple

Answer 12.4
a) It is the innermost layer of the uterine wall consisting of epithelial cells, simple tubular glands and spiral arterioles.

b) It supplies the hormones, oestrogen and progesterone, which are involved in the regulation of the endometrium. It also supplies the oxygen and nutrients needed for growth.
c) This stopped the natural supply of hormones which are produced by the ovaries.
d) Hormone X is oestrogen. This hormone stimulates the proliferation of the endometrial cells causing thickening of the endometrium.
Hormone Y is progesterone. This hormone causes secretion of mucus from the tubular glands which maintains the endometrium.
e) Oestrogen is synthesised by the thecal cells. Progesterone is synthesised by the corpus luteum.
f) The smooth endoplasmic reticulum.
g) The results have relevance to human beings. The two hormones bring about the development and maintenance of the uterine lining for fertilisation. If fertilisation does not occur the fall in levels of these hormones brings about menstruation. The two hormones or progesterone alone can be used for contraceptive pills, preventing ovulation by the maintenance of artificially high levels of the hormones. This prevents the secretion of FSH and no ovulation occurs.

QUESTION, STUDENT ANSWER AND EXAMINER COMMENTS

Question 12.5
The diagram below represents part of a section of a mammalian testis, much magnified.

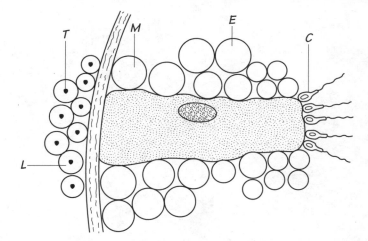

66 All correct 99

Fig. 12.11

a) Rule accurate guidelines and use the relevant letters given below in brackets to label the diagram.
   i)   A region where mitosis is occurring (M)
   ii) A region where meiosis will be taking place (E)
   iii) A region where luteinising hormones will be expected to cause an effect (L)
   iv) A region where a relatively high proportion of mitochondria will be found (often spirally arranged) (C)
   v)   A region where the hormone testosterone is produced (T)   (5)
b) State the name of the largest cell known in the diagram and briefly describe its function.

66 Correct 99

<u>Sertoli cell = Envelops the heads of developing sperm and provides them with</u>
<u>nourishment. Sperm become detached once mature.</u>   (3)

66 Also with support and protection whilst maturing 99

(Total 8 marks)
(ODLE)

NB. Review questions to check your understanding of this chapter can be found at the end of Chapter 13.

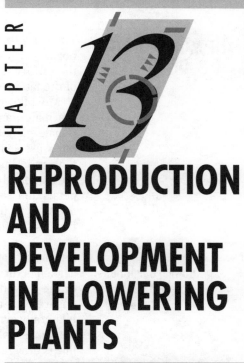

CHAPTER 13: REPRODUCTION AND DEVELOPMENT IN FLOWERING PLANTS

- MORPHOLOGY OF FLOWERS
- POLLINATION AND FERTILISATION
- SEED DEVELOPMENT AND GERMINATION
- PRIMARY AND SECONDARY GROWTH
- ASEXUAL REPRODUCTION
- PRACTICAL WORK

GETTING STARTED

Before starting this topic it will be useful to appreciate the position of the flowering plants in the Kingdom Plantae by referring to Chapter 3, on the diversity of organisms. In discussing the development of the reproductive organs and the growth of the embryo, reference will be made to the processes of **mitosis** and **meiosis**. Development of flowering plants will also involve the action of hormones, which have been referred to in Chapter 9.

PHYLUM ANGIOSPERMOPHYTA

The flowering plants are classified in the Phylum Angiospermophyta, where the **sporophyte** is the persisting plant body, always showing differentiation into **root, stem** and **leaf systems**. The **gametophytes** are very reduced and never exist as independent plants. This group of plants forms the dominant vegetation of the world today and shows a great variety of growth form in the sporophyte.

Herbaceous plants
Plants are **herbaceous** when their aerial parts die down during unfavourable periods. Amongst the herbaceous types are:

- **annuals**, which survive unfavourable periods as **embryos** enclosed in seeds
- **perennials**, which produce flowers and seeds annually and **resting buds** associated with a **perennating** organ containing a food store

Woody plants
Plants are woody when they have a persistent aerial stem protected by an outer covering of **bark**.

The syllabuses vary greatly in their treatment of this topic and it is advisable to check carefully the precise requirements. All syllabuses require a knowledge of the life cycle of a flowering plant, specifying flower structure, formation of gametes, pollination, fertilisation and seed formation. Some require a knowledge of both wind- and insect-pollinated flowers, together with a fairly detailed treatment of outbreeding and pollination mechanisms. There is similar variation in the treatment of embryo development and germination. Some AS- and A-level syllabuses are being developed with Options in Plant Biology, where the treatment of this topic is much detailed and some project work may be required.

CHAPTER 13 REPRODUCTION AND DEVELOPMENT IN FLOWERING PLANTS

THE MORPHOLOGY OF FLOWERS

ESSENTIAL PRINCIPLES

The reproductive structures characteristic of the flowering plants are **flowers**, either borne singly or in a cluster. The arrangement of the flowers on the flower stalk is called the **inflorescence** and several types are distinguishable. They are often characteristic of the families in which they occur.

Most flowers are **hermaphrodite**, the **stamens** forming the **androecium** in which the male gametes develop, and the **carpels** forming the **gynoecium** in which the female gametes develop. In addition to these essential parts of the flower, there are often associated structures, such as the **sepals** and **petals**. The sepals protect the flowers in bud and the petals attract pollinating insects to the plants. The floral parts are arranged in **whorls**, with the carpels in the centre, surrounded by the stamens. The petals and sepals, forming the **perianth**, are around the outside. All these parts are attached to the **receptacle**, which is the expanded head of the flower stalk. In some flowers the floral parts are free, but in others they are joined or fused.

> Drawing a half flower can help display the structure

Drawings of flowers

The structure of flowers is best shown by drawing the different parts as seen in a half of the flower. When preparing to make a half-flower drawing, care must be taken to ensure that a median cut is made in the specimen, particularly in a flower that is bilaterally symmetrical. On the drawing it is conventional to represent any cut edges of petal or sepal by a double line. If the cut has passed between two parts, then a single line is drawn. This is illustrated in Fig. 13.1. Drawings of flowers should be labelled fully and annotated for the functions of the structures. This is a particularly valuable exercise as much information about the pollination mechanism can be incorporated on such drawings. Some syllabuses require a number of different flower types to be studied and half-flower drawings may be set in practical tests.

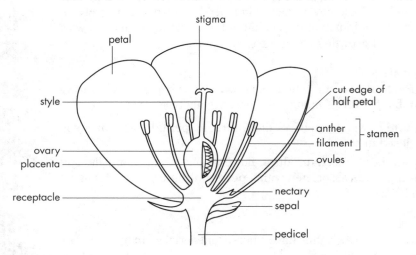

Fig. 13.1 A generalised flower: half-flower diagram

THE ANDROECIUM

The androecium consists of the **stamens**. Each stamen is made up of an **anther** and a **filament**.

- The filament holds the anther in a suitable position within the flower. In wind-pollinated flowers, the filaments are attached to the middle of the anther and the mature stamens hang outside the flower. As well as supporting the anther, the filament contains vascular tissue, which transports food materials necessary for the formation of the pollen grains.

- The **anther** is usually made up of four **pollen sacs** arranged in two pairs, side by side.

Pollen grain development

In the early stages of development, each **pollen sac** contains a mass of **microspore mother cells**, surrounded by a nutritive layer called the **tapetum**. Each microspore mother cell undergoes **meiosis** to give a tetrad (four) of **haploid** cells. Each of these cells develops into a **pollen grain**. A thick outer wall, the **exine**, and a thin inner wall, the **intine**, are secreted. The exine may be pitted or sculptured, depending on the species. Inside the pollen grain the haploid nucleus undergoes mitosis to produce a generative nucleus and a **pollen tube** nucleus. The generative nucleus represents the male **gametophyte** and will later give rise to the two **male gamete nuclei**. It is important to realise the great reduction in the gametophyte and to note that there are no vegetative cells involved. As the pollen grains mature, changes take place in the anthers. The cells of the tapetum shrink and degenerate, and fibrous layers develop beneath the epidermis. When the pollen is ripe, the outer layers of the anthers dry out and tensions are set up in lateral grooves. Eventually dehiscence occurs and the edges of the pollen sacs curl away, exposing the pollen grains.

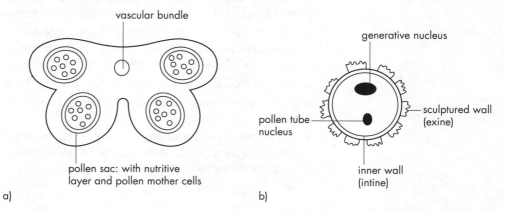

Fig. 13.2 a) Anther and b) pollen grain cross-sections

THE GYNOECIUM

The **gynoecium** is made up of one or more **carpels**. Each **carpel** is a closed structure, inside which one or more **ovules** develops. Each ovule develops from the **nucellus** and becomes surrounded by two **integuments**. The lower part of the carpel, which surrounds the ovules, is called the **ovary** and bears at its apex a stalk-like structure, the **style**. This ends in a receptive surface, the **stigma**.

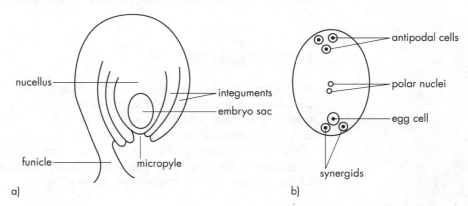

Fig. 13.3 a) Ovule and b) embryo sac

Ovule development

In the early stages of the development of an **ovule**, the **nucellus** arises as a tiny lump of tissue growing out from the **placenta** of the carpel. At its apex a **megaspore mother cell** develops and undergoes **meiosis** to form four **haploid megaspores**. Normally only one of these cells continues to develop; the rest abort. The single **megaspore** gets bigger, gaining its nutrition from the nucellus tissue. It undergoes three successive **mitotic** divisions and forms an **embryo sac** containing eight **haploid** nuclei. This embryo sac represents the female **gametophyte**, a very reduced structure when compared to the gametophytes of the mosses and ferns. As the nucellus enlarges, one or more **integuments** grow up from around the base, but they do not completely surround it. A small pore or opening called the **micropyle** is left at the apex of the ovule. In the embryo sac the eight nuclei, each associated with some cytoplasm, are arranged in a definite pattern. Three cells are situated at the **micropylar** end of the ovule, making up what is referred to as the **egg apparatus**. The central cell is the **female gamete** or **egg cell**, and the surrounding cells are called **synergids**. At the opposite, or **chalazal**, end is another group of three cells called the **antipodal cells**. The remaining two nuclei, called the **polar nuclei**, migrate to the centre of the embryo sac. They may remain separate at this stage, but eventually fuse to give rise to a **central fusion nucleus**, which is diploid.

Variations

This type of development is one of several kinds which are known in flowering plants, and variations occur in the details, particularly with respect to which megaspore survives and to the number of nuclei eventually present in the embryo sac.

- The egg nucleus usually remains haploid, but the chalazal polar nucleus may be diploid, so that the central fusion nucleus become **triploid**.
- The carpels may be separate (**apocarpous condition**) or fused together (**syncarpous condition**), the fusion occurring laterally and resulting in the formation of one or more internal cavities called **loculi**.
- The position in which the ovules develop may be **marginal**, **axile** or **central**.
- In the apocarpous condition each carpel has its own style, with the stigma at the apex.
- In the syncarpous condition the styles may be separate, but more usually they are united to form a single structure bearing lobed stigmas, the number of lobes corresponding to the number of carpels.

POLLINATION AND FERTILISATION

Pollination is necessary in order that the pollen grains, containing the male gametes, are brought into contact with the gynoecium so that fertilisation can be achieved. This means that pollen grains must be transferred from the ripe anther to the receptive stigma.

SELF-POLLINATION

In some species self-pollination occurs and pollen from the anthers of a flower need only be transferred to the stigma of the same flower, or another flower on the same plant.

CROSS-POLLINATION

In a large number of species, cross-pollination is more normal, where pollen is transferred from the anthers of one flower to the stigma of another flower on another plant of the same species. There are a variety of mechanisms which either ensure cross-pollination or prevent self-pollination, so ensuring **outbreeding**. Anthers and stigma may mature at different times, they may be at different levels in the flower or there may be separate male and female flowers on different plants.
There are two main methods of pollination:

- by wind
- by animal vectors such as insects, birds or mammals

WIND-POLLINATION

Wind-pollinated flowers tend to be inconspicuous, with small petals and sepals, lacking scent or nectaries. The stamens are either very numerous, or there are large anthers present, producing masses of light, smooth pollen. The filaments are very long and dangle outside the flower so that the pollen is easily dispersed by air currents. The stigmas are long and feathery, providing a large surface area to trap the pollen. They also often hang outside the flower thereby catching pollen in air currents (see Fig. 13.4).

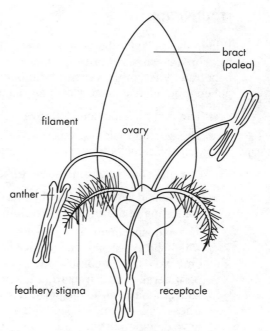

Fig. 13.4 Grass flower

- Trees such as oak and elm have wind-pollinated flowers which appear before the foliage leaves.
- Herbaceous plants such as grasses are also wind-pollinated and their inflorescences are borne above the surrounding leaves.

INSECT-POLLINATION

Insect-pollinated flowers are large and brightly coloured to attract the insects. They have coloured, often shiny, petals and produce a scent and nectar. The scent attracts the insects and the nectar is collected for food. The stamens are fewer in number than in the wind-pollinated flowers and they are fixed inside the flower, where insects must brush past them in order to reach the nectar. The pollen grains are large, often with elaborately sculptured walls, which assist in enabling them to stick to the bodies of the insects. The stigmas are often lobed and remain within the flower. Insect-pollinated flowers have highly specialised pollination mechanisms, adapted to the type of visiting insects (see Fig. 13.5).

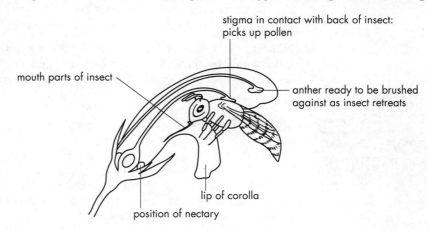

Fig. 13.5 Insect visiting a flower

SUCCESSFUL POLLINATION

Successful pollination results in **compatible** pollen grains reaching the stigma. if the pollen grains are incompatible, then they will not germinate, or if they do then growth will soon cease. The stigma produces a sugary solution in which the pollen grains germinate, producing a fine **germ tube** or **pollen tube**. This grows down into the tissues of the style, secreting enzymes, which digest some of the cells in order to gain nutrients. The **pollen tube nucleus** is situated near the top of the tube, with the two **male nuclei** close behind.

FERTILISATION

When the pollen tube reaches the **loculus** of the carpel, it grows round to the **micropyle**, passes into the ovule and then to the embryo sac. As the pollen tube penetrates the embryo sac, the tip opens and the two male gamete nuclei enter, the pollen tube nucleus having disintegrated earlier.

- The first male nucleus fuses with the **female**, or **egg nucleus**, to form a **zygote**.
- The second male nucleus passes further into the embryo sac and fuses with the two polar nuclei to form a **polyploid endosperm nucleus**.

> Double fertilisation is a unique feature of flowering plants

These events constitute **double fertilisation**, a unique feature of the flowering plants.

- The diploid zygote nucleus begins to divide by **mitosis** and eventually an embryo, consisting of a **plumule**, a **radicle** and one or two **cotyledons**, is formed.
- The endosperm nucleus also divides mitotically, forming a large number of nuclei, which lie freely in the cytoplasm of the embryo sac. The central part of the embryo sac becomes vacuolated and these nuclei arrange themselves peripherally and are surrounded by cellulose cell walls. Division of cells continues and gradually an endosperm tissue is built up around the developing embryo. These cells become filled with stored food materials, providing reserves for the developing embryo.

Changes

Changes take place in the **integuments**, which become fused together and thickened with **lignin** and **cutin**. This layer is now the **testa** and its toughness provides good protection for the dormant seed. During the later stages of development the seed dries out, losing as much as 90% of its water, and in this state it is dispersed from the plant.

- Changes also take place in the **carpel wall**, the **pericarp**, and often in the **receptacle**, forming the **fruit** from which the seeds are eventually dispersed.

Seed dispersal

In many cases, the pericarp becomes dry and the seeds are either dispersed by wind or by the fruits sticking to the coats of passing animals. Sometimes the fruit is explosive, as in lupin, and the seeds are shot out with considerable force as the fruit dries.

- The pericarp may become fleshy and thus palatable to animals, who drop the seeds at great distances away from the parent plant. In some cases the seeds pass through the gut of the animals before being dispersed.

Not many syllabuses require a comprehensive knowledge of fruits and their dispersal, but it would be sensible to appreciate the general principles and to know the dispersal mechanisms of the flowers which you have studied in detail.

SEED DEVELOPMENT AND GERMINATION

SEED DEVELOPMENT

As embryo development proceeds the endosperm tissue gradually disappears, the nutrients passing into the embryo. In some seeds the endosperm tissue disappears altogether and the **cotyledons** become swollen with the stored food (e.g. in peas and broad beans), but in others the endosperm remains as part of the seed and the cotyledons stay thin (e.g. in castor oil seeds).

GERMINATION

Germination of the seed usually occurs after a period of dormancy, and then only if the right conditions prevail. These conditions are:

- a suitable temperature
- sufficient moisture
- sufficient oxygen

> Note the conditions necessary for germination

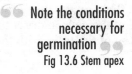
Fig 13.6 Stem apex

Water is taken up rapidly by the seed in the initial stages, causing the tissues to swell as well as mobilising the enzymes that will make the stored food available for the growth of the embryo. The **testa**, or seed coat, will rupture as the **radicle** pushes its way through. The radicle will grow downwards, and the **plumule** upwards, under the influence of gravity.

Epigeal germination
In epigeal germination the cotyledons rise above the ground as a result of the elongation of the **hypocotyl**; the region of the embryo above the radicle and below the cotyledon stalks. The cotyledons develop chlorophyll and function as the first leaves before the plumule has developed fully, e.g. sunflowers.

Hypogeal germination
In hypogeal germination the cotyledons remain below the ground throughout. The region of the embryo which elongates is the **epicotyl**, the portion below the plumule and above the cotyledon stalks. This brings the plumule above the ground and the first leaves quickly expand and begin to photosynthesise e.g. in pea.

These are the two main types of germination.

DEVELOPMENT OF THE SHOOT AND ROOT

Apical meristems
At the tip of the shoot and the root are regions of actively dividing cells called apical meristems. The cells are dividing **mitotically** and observations show that the cells at the surface are arranged in rows, while those beneath are more haphazardly arranged. The cells at the surface make up the **tunica** and divide so that new cell walls are formed at right angles to the surface, increasing the area. The cells below form the **corpus** and divisions here can be in different planes, increasing the bulk of the organ by broadening and lengthening it.

- The tunica gives rise to the **epidermal tissues.**
- The corpus gives rise to the **cortex, medulla** and **vascular tissues**.

CELL EXPANSION

During cell expansion, the new cells formed develop small **vacuoles** in the cytoplasm, and these take up water by osmosis. The vacuoles enlarge and eventually coalesce into one large **sap vacuole**. The cell is able to expand at this stage because the cellulose cell wall is thin and stretchable.

CELL DIFFERENTIATION

Cell differentiation begins with the thickening of the cellulose walls, involving changes in shape due to the uneven deposition of the new cellulose.

- If the **cellulose microfibrils** are laid down evenly, then the cell will become spherical.
- If the deposition is uneven, then the cell may become long and narrow.

CELL DEPOSITS

- **Cellulose** may be deposited in the corners of the cells, forming typical **collenchyma** tissue.
- **Lignin** may be deposited amongst the cellulose microfibrils, resulting in the formation of **woody tissue**. In this case the walls of the cells are rendered impermeable to nutrients and the cell contents die. This leaves the cells empty with hollow **lumens**, typical of **vessels, tracheids** and **fibres**.

STEM APEX

It is possible to distinguish three clear zones behind the apex of the stem or root:

- the zone of cell division
- the zone of cell expansion
- the zone of cell differentiation

The apical meristem of the stem gives rise to **superficial leaf primordia**, which appear as tiny bumps at the apex. These grow and develop around the apex, forming an **apical bud**, and protecting the meristematic tissue. In the angle formed by the leaves and the main stem, **lateral**, or **axillary**, buds form. Their origin is superficial, or **exogenous**. These have a similar structure to the apical bud and are potentially capable of growing into lateral branches. Leaves arise at the **nodes** on the stem, and the portion of stem between two nodes is called the **internode** (see Fig. 13.6).

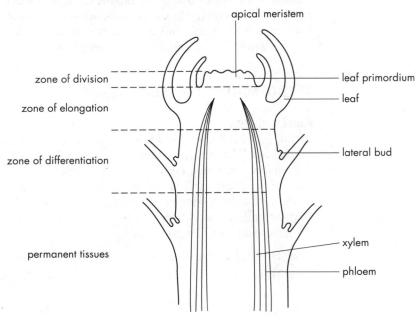

ROOT APEX

Lateral roots arise from small groups of cells adjacent to a **protoxylem** group in the **pericycle** of the root. A new apical meristem develops at the tip of the lateral root and it grows out through the tissues of the **cortex**. Thus the lateral roots arise from within the root, or **endogenously**.

PRIMARY AND SECONDARY GROWTH

Apical growth of the root and the stem results in increase in **length** of these organs. Increase in **girth**, which occurs in **perennial** trees and shrubs, is achieved by the activity of a **lateral meristem**: the **vascular cambium**.

VASCULAR CAMBIUM

This is a tissue in which the cells have retained their ability to divide, and it is located between the xylem and the phloem. The divisions are **mitotic**, new xylem tissue being formed to the inside and new phloem tissue to the outside. **Radial divisions** of the cambium cells increase the circumference of the tissue to keep pace with the increasing girth. Two types of vascular **cambial cells** occur:

- the **fusiform** initials, long narrow cells which give rise to the new vascular tissues
- the **ray** initials, almost spherical in shape, which produce **parenchyma** cells, forming **rays** between the phloem and the xylem. These rays are important for the lateral transport of materials in the woody tissue.

SECONDARY GROWTH

The activity of the vascular cambium results in secondary growth, the new tissues being called **secondary xylem** and **secondary phloem** to distinguish them from the tissues of the primary plant body.

CORK CAMBIUM

The increase in girth imposes a strain on the epidermal tissues and these are replaced by the activity of another lateral meristem, the cork cambium. This arises just below the epidermis, forming **secondary cortex** cells to the inside and a layer of **cork** cells to the outside. The walls of the cork cells become impregnated with **suberin**, a fatty material that makes them impervious to water and gases. At intervals in the **bark** which is formed by the cork cambium there are groups of loose cells through which respiratory gases can diffuse. These groups of cells are called **lenticels**, and reference has already been made to them in Chapter 8 dealing with gas exchange.

GROWTH RINGS

The secondary growth which has been described occurs during the growing season each year, adding characteristic rings of tissue to the stems and roots of woody plants. These rings can be seen with the naked eye where logs have been sawn across; it is possible to estimate the age of a tree by counting these **annual rings**. They also give 'figure' to timber used in the manufacture of furniture and ornaments.

Although only secondary growth in stems has been described here, similar events take place in roots, but most syllabuses do not require a knowledge of this.

ASEXUAL REPRODUCTION

66 Check your syllabus 99

Many syllabuses require a general understanding of asexual reproduction in plants, but most boards do specify 'in outline only'. A more detailed knowledge is required for Options in Plant Biology, particularly when applied to the commercial applications. It is wise to check what you need to know.

PROPAGULE FORMATION

Asexual reproduction involves the formation of **propagules**, involving **mitotic divisions**. These can then become detached from the parent plant and eventually exist on their own. The propagules may be formed from:

- modified stems – runners and stem tubers
- roots and leaves

Perennating organs
Sometimes large quantities of food are stored and the propagule may also be a perennating organ. This state of affairs is characteristic of the **herbaceous perennials**.

Bulb formation
In bulb formation food is stored in swollen leaf bases for the growth of next year's terminal bud. Lateral buds often develop and can become detached. They are capable of growing independently of the parent plant, thus displaying **vegetative propagation** as well as **perennation**.

Propagation
Propagules are **genetically identical** to their parent plants, so this form of propagation is often used by gardeners and horticulturalists. There are also artificial means of propagation, which involve the taking of **cuttings** from stems or root, and **budding** or **grafting**. Reference is made to these topics in some sections of both A- and AS-level syllabuses which are being developed.

CHAPTER 13 REPRODUCTION AND DEVELOPMENT IN FLOWERING PLANTS

PRACTICAL WORK

There is not a great deal of experimental practical work involved in this topic, but observations of flowers, seed structure and development will all help in gaining an understanding of the basic facts. Students will find it valuable to make accurate drawings of flowers, annotating these for their methods of pollination. Pollen grains can be observed easily using the light microscope, and it is possible to get them to produce pollen tubes if kept in a suitable concentration of sucrose.

Further reading
Street H E and Opik H, 1984, *The Physiology of Flowering Plants*. Arnold
Clegg C J and Cox G, 1978, *Anatomy and Activities of Plants*. J. Murray

EXAMINATION QUESTIONS AND ANSWERS

QUESTIONS

Question 13.1
a) i) Draw a large, labelled diagram to show the structure of a named cereal grain as seen in longitudinal section. (5)
   ii) Name the main storage compound in the cereal grain you have drawn. (1)
b) i) Germination is triggered off by the imbibition of water by the seed. Explain how this affects the embryo and results in its subsequent growth. (4)
   ii) State **two** other conditions which are necessary for successful germination of seeds. (2)

(Total 12 marks)
(ULEAC AS-level)

Question 13.2
The diagrams below show stages in the life cycle of a flowering plant.

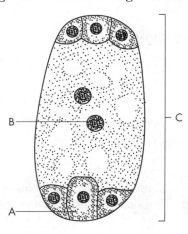
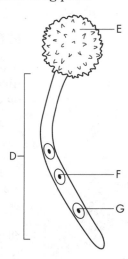

Fig. 13.7

a) Complete the table below by choosing the letter from the diagrams which refers to each of the stages given. (4)

STAGE IN LIFE CYCLE	LETTER
Female gametophyte	
Tube nucleus	
Female gamete	
Male gamete	

b) i) State **one** function of D.
   ii) How has structure D enabled flowering plants to adapt to terrestrial life? *(3)*
c) Comment on the surface structure of E. *(2)*
d) Suggest **two** ways in which self-fertilisation may be avoided in flowering plants. *(2)*

(Total 11 marks)
(ULEAC)

Question 13.3
Compare the process of sexual reproduction in a flowering plant and in a mammal.
(20)
(Total 20 marks)
(ULEAC)

Question 13.4
The diagram below shows a vertical section through a flower from the family Papilionaceae (Leguminosae).

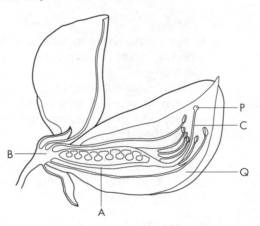

Fig. 13.8

a) i) Identify structures A, B and C. *(3)*
   ii) Indicate on the diagram by means of a clear arrow and a label M, **one** structure in which meiosis occurs. *(1)*
b) State precisely how structures P and Q are involved in pollination. *(2)*

(Total 6 marks)
(ULEAC)

OUTLINE ANSWERS

Answer 13.1
a) i) Credit would be given for the general size and proportions of the drawing, together with appropriate detail of the aleurone layer and the embryo. Labels should include: plumule, radicle, embryo, cotyledon/scutellum, endosperm, pericarp.
   ii) Starch
b) i) The embryo absorbs water and secretes gibberellins, which cause the aleurone layer of the seed to produce enzymes. Amylase will catalyse the breakdown of starch to maltose and maltase will convert maltose to glucose.
   ii) Suitable temperature; oxygen

Answer 13.2
a) Female gametophyte is C
   Tube nucleus is G
   Female gamete is A
   Male gamete is F
b) i) Provides passage for male gametes.
   ii) Means that fertilisation can be independent of water, and the gametes are protected from drying up.
c) Surface is rough or spiky, so can stick to insects or to the surface of the stigma. Specific shapes of some pollen grains can be recognised by stigma.

d) Protandry (male parts mature before female parts)
   Protogyny (female parts mature before male)
   Sexes borne on separate plants

Answer 13.3
This is an open-ended topic and will require a great deal of organisation of the material if a high mark is to be gained. The question says 'compare', so students must remember to mention similarities as well as differences. In addition it is not a good idea to write all about the flowering plant and then all about the mammal; try to make points about each all the way through.

Similarities
Similarities include reference to fertilisation essentially being the same in both – fusion of male and female gametes to give a zygote. Both types of male gametes are produced in large numbers, there are fewer gametes in both and all gametes are haploid. In both the zygote develops inside the female in a protected environment.

Differences
General differences include separate sexes in animals, alternation of generations in flowering plants, active male gametes and necessity for fluid for male gametes in mammals.

The process of sexual reproduction can then be divided up into gamete production, the way in which gametes are transferred from male to female in both, followed by development of the embryo in both. There are differences in all these areas and the details should be included.

In this type of essay, it would be a waste of time to draw too many diagrams, and certainly diagrams of the reproductive systems would be inappropriate. Marks are only gained from diagrams if there are annotations or they show something clearly relevant to the question set.

Answer 13.4
a) i) A Ovary wall (NB: if the whole ovary was required, a bracket label would be used.)
      B Receptacle
      C Style
   ii) M should be arrowed to either an ovule or an anther.
b) A bee lands on and depresses Q to expose P. P brushes against the bee's body and gets pollen from the previous flower the bee visited.

QUESTION, STUDENT ANSWER AND EXAMINER COMMENTS

Question 13.5
The diagram on p. 179 (Fig. 13.9) represents a vertical section through the carpel and ovule of a flower such as buttercup. Pollination has occurred.

a) Give the names of the structures labelled A-H in the spaces below.
   A style   E funicle
   B cavity of ovary F nucellus
   C ovary wall   G micropyle
   D integuments   H receptacle   (4)

> All these structures have been correctly named. If there is some confusion, it may be due to the lines the student has drawn in to answer another part of the question

b) Suppose that fertilisation has just occurred. On the diagram draw in clearly and label the following structures:
   i) the zygote cell
   ii) two synergid cells
   iii) the endosperm nucleus
   iv) the antipodal cells
   v) the pollen tube   (5)

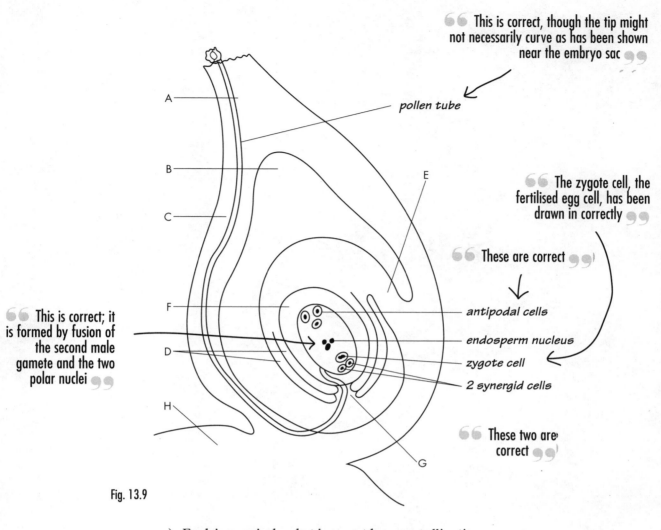

Fig. 13.9

> This is correct, though the tip might not necessarily curve as has been shown near the embryo sac

> The zygote cell, the fertilised egg cell, has been drawn in correctly

> These are correct

> This is correct; it is formed by fusion of the second male gamete and the two polar nuclei

> These two are correct

c) Explain concisely what is meant by *cross-pollination*.

> Good definition of cross-pollination

When the pollen grains of one flower are carried to the stigma of a flower of another plant of the same species. (2)

d) How is endosperm formed?

One of the male nuclei from the pollen tube fuses with both polar nuclei in the embryo sac to form a triploid nucleus this divides mitotically and the cells become a food storage tissue. (2)

> Correct answer

e) What part could *chemotropism* play in the processes of fertilisation in flowering plants?

The pollen tube tip could show chemotropic response by growing directly towards chemicals released by the embryo sac. This ensures the male nuclei reach the correct place. (2)

> This is right

f) What is the significance of the spiny nature of the pollen grain wall?

The spines make it easier for the pollen grain to stick to insects (etc.) which can carry it to another flower allowing pollination. (1)

> Correct answer

g) What do the structures labelled D on the diagram eventually form?

Seed coat = testa (1)

> Correct answer

> A good answer, showing understanding of the terms and a knowledge of the process of fertilisation.

(Total 17 marks)
(ODLE)

REVIEW SHEET

1 Complete the labels in each of the following diagrams.

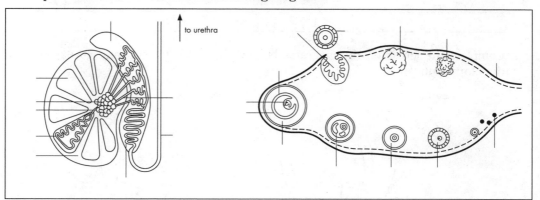

2 Describe the sequence of events involved in **gametogenesis**.
_____
_____
_____
_____

3 Describe the sequence of events involved in the **formation of ova**.
_____
_____
_____
_____

4 Complete the labels to the following diagram. Then use the diagram to explain **fetal circulation**.

_____
_____
_____
_____
_____
_____
_____
_____
_____
_____
_____
_____
_____

5 Briefly describe the key functions of the **placenta**.
_____
_____
_____
_____

6 Explain the stages involved in **pollen grain development**.
___
___
___
___
___

7 Complete the labels on the diagrams below. Then use the diagrams to explain **ovule development**.

ovule   embryo sac

___
___
___
___

8 Write briefly on each of the following.
   a) Cross pollination: ___
   ___

   b) Wind pollination: ___
   ___

   c) Insect pollination: ___
   ___

9 Describe the key stages involved in **double fertilisation** in flowering plants.
___
___
___
___

10 Write briefly on each of the following.
   a) Epigeal germination: ___
   ___
   ___

   b) Hypogeal germination: ___
   ___
   ___

CHAPTER 14
SUPPORT, MOVEMENT AND LOCOMOTION

GETTING STARTED

Support, movement and locomotion are all interconnected and have inter-relationships with other aspects of anatomy and physiology. Many physiological functions depend upon support: for example, the placing of leaves in the best position for photosynthesis; the support of animal organs for respiration. Similarly, animals are dependent on movement and/or locomotion for food gathering and reproduction. In mammals movement and locomotion are controlled by nerves; in plants movement is governed by hormones and growth substances. Questions often span all these areas. Also, in plants support and transport are intimately connected as often the same cells are concerned in both.

Questions often incorporate sections on relationships between tissues and organs or comparisons of endoskeletons, exoskeletons and hydroskeletons. You may need to know the positions, origins and insertions of the main muscles moving the elbow and knee joints in a mammal. Diagrams often feature in questions on these topics and it would be wise to be able to draw and label diagrams of the main types of cells involved and the arrangement of supporting tissues in stems, leaves and roots. A practical study of support cells and tissues is important and you should be familiar with low- and high-power microscopy of the relevant tissues, including the morphology of the major bones in mammals, features of the skeleton in insects and perhaps of other arthropods.

- **MAMMALIAN SKELETONS**
- **LOCOMOTION IN VERTEBRATES**
- **EXOSKELETON AND MUSCLES IN INSECTS**
- **HYDROSTATIC SKELETONS**
- **AMOEBOID MOVEMENT**
- **CILIA AND FLAGELLA**
- **SUPPORT TISSUES IN PLANTS**
- **PRACTICAL WORK**

CHAPTER 14 SUPPORT, MOVEMENT AND LOCOMOTION

MAMMALIAN SKELETONS

ESSENTIAL PRINCIPLES

A mammalian skeleton is divided into:
- the **axial** skeleton, consisting of the skull and vertebral column
- the **appendicular** skeleton, made up of the pectoral and pelvic girdles with the fore and hind limbs

THE SKULL

The skull consists of two major parts:
- the **cranium**
- the **jaws**

The **cranium** consists of a series of flat bones which are joined by fixed joints, **sutures**. Its main function is to protect the brain, eyes, middle ear and olfactory organs. Its other function is to support the jaws (see Fig. 14.1).

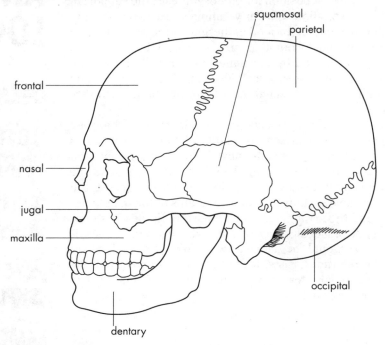

Fig. 14.1 Human skull

THE VERTEBRAL COLUMN

The vertebral column is made up of five different types of vertebrae all of which have the same basic structure as shown in Fig. 14.2. If your syllabus specifies a knowledge of detail of the different vertebrae, refer to the standard texts for these.

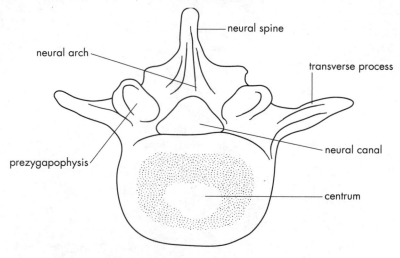

Fig. 14.2 A generalised vertebra

LIMB GIRDLES

Limb girdles serve the function of attaching limbs to the vertebral column. They differ in structure and function depending on their different purposes in locomotion (see Fig. 14.3).

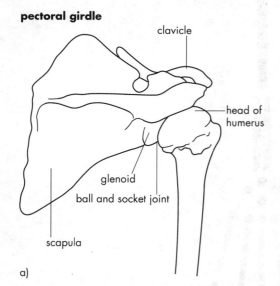
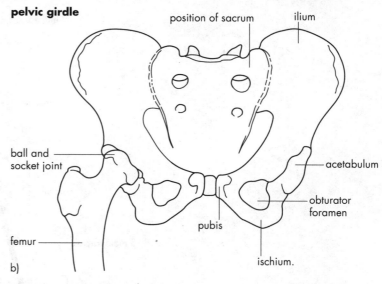

Fig. 14.3 a) Human pectoral and b) pelvic girdles

- The **pectoral girdle** consists of two identical but separate halves, each suspended from the thoracic spine by ligaments. It is able to move through various angles and although strong enough to support the animal at rest, its main function is to act as a shock absorber when animals land after a jump and when running fast on four limbs.
- The **clavicle**, in bipedal animals, acts as a fulcrum for sideways movements of the fore limbs. This is particularly important in the higher primates allowing **brachiation** and the use of tools.
- The **pelvic girdle** also consists of two halves but these are fused, both together and to the sacral spine. This gives a strong pivot for the hind-limb muscles to work against in producing the power for movement.

LIMB PLANS

The limbs of mammals all have the same basic pattern, the **pentadactyl** limb plan. This becomes modified in different species to fulfil different functions, which is an example of **adaptive radiation**. The main bones in a limb consist of a shaft, which forms a lever, and ends, which are shaped to form the joints between the bones. At the ends of the bones there are also various ridges, which form points of attachment for the skeletal muscles. Apart from its skeletal functions, the bony skeleton is involved in the production of blood cells by the **red bone marrow** and in the maintenance of constant levels of calcium and phosphorus in the bloodstream (see Fig. 14.4, p. 186).

JOINTS

Whenever two bones meet a **joint** is formed. In bony skeletons three different types of joint are found.

Immovable joints
Immovable joints are found in the skull, sternum and pelvis and are formed by a thin layer of fibrous connective tissue holding the bones firmly in position. These joints help the bones to give protection to delicate organs and provide strength and support.

Partially movable joints
Partially movable joints have cartilaginous pads between the bones which allow a certain amount of movement as the bones can glide over each other or rotate in relation to each other. Examples of these are the wrist, ankle and joints between the vertebrae (see Fig. 14.5, p. 186). Although the movement of each bone is limited, overall a wide range of movements is possible.

CHAPTER 14 SUPPORT, MOVEMENT AND LOCOMOTION

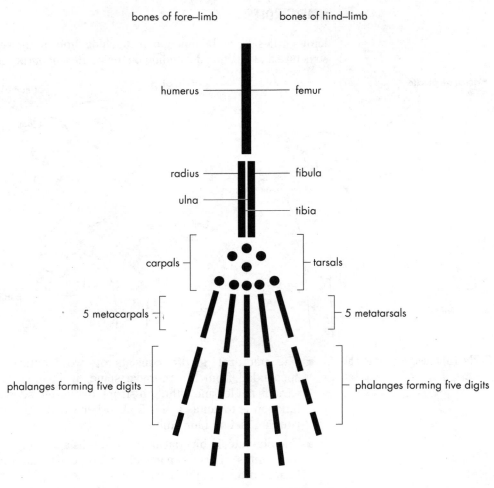

Fig. 14.4 Pentadactyl limb plan

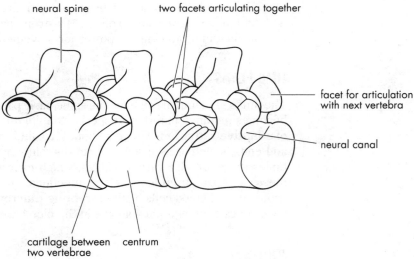

Fig. 14.5 A sliding joint

Synovial joints

Synovial joints are characterised by articular cartilages on the ends of the bones being separated from each other by a capsule filled with **synovial fluid** (see Fig. 14.6). Synovial joints take the form of hinge joints at the knee and elbow, with movement limited to one plane, and ball and socket joints at the hip and shoulder which can move in all planes and also rotate.

BONE

Bone consists of a matrix of collagen fibres which becomes impregnated with calcium salts. Compact bone consists of a series of **haversian canals** parallel to the long axis of the bone. **Lamellae** containing **osteocytes** are laid down in concentric circles round

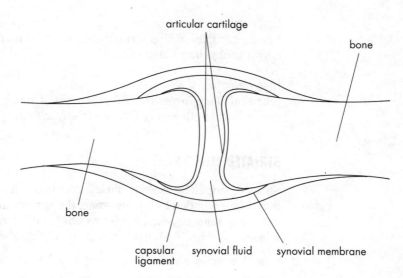

Fig. 14.6 A synovial joint

> Note that lacunae but not bone cells can be seen in sections of decalcified bone

them. In the ends of the bones the bone tissue is laid down as struts, **trabeculae**, which follow the lines of compression force to give strength to the bone. Usually bone is laid down to replace embryonic cartilage (see Fig. 14.7). Chondroblasts in the cartilage lay down calcium compounds. They are then replaced by amoeboid **osteoclasts**, which erode away the calcified cartilage. This is replaced with bone laid down by osteoblasts, which become enclosed in **lacunae** in the matrix. The shafts of long bones are hollow, with an outer cylinder of hard bone.

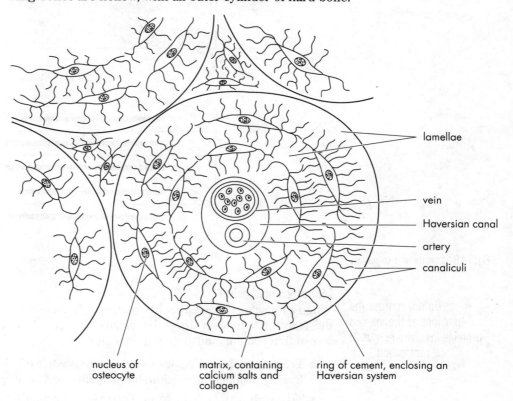

Fig. 14.7 Haversian systems in bone

CARTILAGE

Three types of cartilage are present in vertebrates.

Hyaline cartilage
Hyaline cartilage is found covering the articular surfaces of bones; it joins the ribs to the sternum. It consists of a matrix of chondrin, secreted by **chonodroblasts** found in pairs or clusters in the matrix.

Elastic cartilage
Elastic cartilage forms structures such as the epiglottis and ear pinna and contains a network of yellow elastic fibres.

Fibro-cartilage
Fibro-cartilage forms the intervertebral discs; it contains many white collagen fibres, giving it strength and resistance to stretching.

STRIATED MUSCLE

Striated, striped or skeletal muscle consists of fibres made up of many **myofibrils**. Each fibre is surrounded by a membrane, the **sarcolemma**, which contains several nuclei and has many mitochondria. Each myofibril is divided into **sarcomeres** by cross-partitions, Z lines (see Fig. 14.8). In each sarcomere there are two types of protein fibres:

- thin **actin** strands projecting into the sarcomere from the Z lines
- thicker **myosin** strands between them, which are fixed in position by molecular bonds

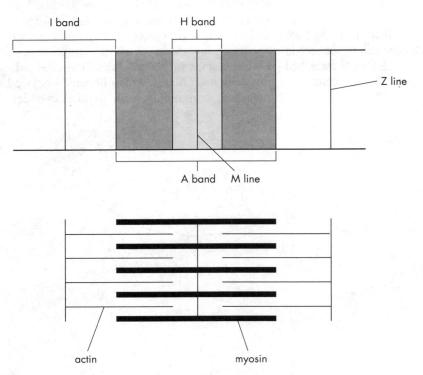

Fig. 14.8 Structure of a sarcomere

> Do not confuse the functions of the various proteins involved in muscle contraction

Contraction mechanism
When a muscle contracts, each sarcomere shortens when the actin strands glide along the stationary myosin fibres by means of a ratchet mechanism. Two proteins are involved in starting and stopping the contraction.

- **Tropomyosin** serves to switch on or switch off the contractile mechanism by freeing or blocking calcium-binding sites on the actin filament.
- **Troponin** consists of three sub-units: one binds troponin to tropomyosin; one provides calcium-binding sites; and one can inhibit the actin/myosin interaction if calcium ions are absent.

Calcium ions enter the sarcomere via the **T system** of microtubules in the cytoplasm (**sarcoplasm**). The energy for the attachment of actin to myosin is provided by ATP. The ATPase activity is in turn controlled by the level of calcium ions in the fibril. The release of energy from ATP accompanies cross-bridge formation and changes the angle of the cross-bridges so that the myosin pulls the actin towards the centre of the sarcomere. The ATP is produced by aerobic respiration in the mitochondria, associated with the myofibrils. Reserves of oxygen are held in the muscles in combination with the respiratory pigment **myoglobin**.

Oxygen debt

When oxygen is in short supply because muscles are in active use, energy can be supplied by the conversion of respiratory substrate to **lactic acid**. This acts as a muscle poison, causing fatigue, and must be removed as soon as possible. The production of lactic acid causes the **oxygen debt**. When oxygen becomes available this debt is repaid and the lactic acid is oxidised. This process in turn provides ATP. If required, you will find further detail of the excitation-coupling-contraction mechanisms of muscle in the standard textbooks.

LOCOMOTION IN VERTEBRATES

> Remember that muscles must be in pairs or groups as they only work by contraction – muscles can only pull not push

Muscles

Locomotion is brought about by the movement of the long bone lever systems at the various joints. At each joint, groups of muscles act. These are antagonistic **flexor** and **extensor** muscles with subsidiary muscles which steady the joint.

Nerve muscle junctions

The contraction of the various muscles is controlled by nerve fibres which innervate the muscle fibres. Nerve muscle junctions are similar to synapses between nerve fibres and use neurotransmitters which depolarise the membrane of the muscle fibre to stimulate its contraction. Because joints are moved by antagonistic muscles, as well as stimulatory fibres, to each muscle there are also inhibitory fibres so that when one muscle contracts its antagonist is inhibited. Whenever a muscle is stretched, stretch receptors or **muscle spindles** are stimulated, initiating a reflex which will cause the muscle to contract. During movement, these reflexes are inhibited but when an animal is at rest they function to produce slight alternate contractions in the antagonistic muscles, maintaining the animal in position. The muscles are thus kept in a slightly contracted state and they are said to have **tone**.

Amongst the vertebrates, there are many different types of locomotion. The shapes of limb bones and the size and organisation of the muscles depends on the locomotion type. Check with your own syllabus which types of locomotion are included and refer to the standard texts for the detail you need.

EXOSKELETON AND MUSCLES IN INSECTS

Exoskeletons are characteristic of the phylum Arthropoda. They are secreted by the epidermal cells and consist mainly of **chitin**, a light, tough and flexible skeletal material. Chitin is composed of **glucosamine** polymers, sometimes strengthened by hardened proteins or in some aquatic crustaceans by calcium carbonate. The chitin remains unmodified in the joints between parts of the skeleton to give flexibility. At

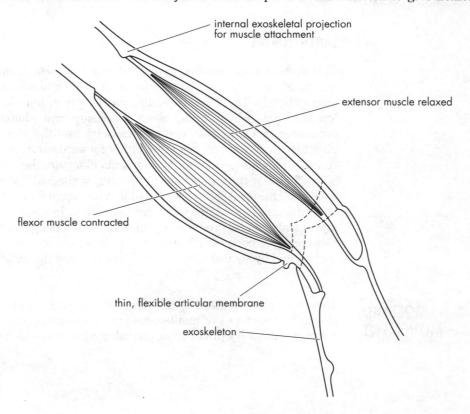

Fig. 14.9 Exoskeleton and associated muscles in an insect limb

either side of a joint there are inward projections, **apodemes**, forming points of attachment for muscles which move the plates on the body to produce respiratory movements and flex or extend the limbs to allow locomotion (see Fig. 14.9). Chitin itself is permeable to water and so an outer waxy **epicuticle** is produced by gland cells in the epidermis to give waterproofing.

Although an exoskeleton is very efficient, it is only suitable for small animals. There is also a problem with growth. To allow growth the exoskeleton must be shed and replaced with a new one which is soft and extensible. Growth takes place before the new skeleton hardens and then stops until the next moult or **ecdysis**.

WALKING IN INSECTS

Insect legs are joined to the thorax, one pair to each segment, by ball and socket joints. The other joints are hinge joints. The limbs are moved by antagonistic flexor and extensor muscles. When moving, the legs work as two sets of three. The first leg on one side pulls whilst the third leg pushes, the second leg on the other side also helping the forward movement. During this process the other three legs remain on the ground, supporting the body. Then the two sets of legs reverse roles and the alternation continues whilst the insect is moving.

FLIGHT IN INSECTS

Insect wings are flattened extensions of the exoskeleton supported by hollow veins; they are attached to the dorsal and lateral plates of the thoracic skeleton.

> Wings of insects are moved by muscles external to their own structure, contrast this with the structure of bird wings

- Large-winged insects such as butterflies and locusts have muscles attached to the bases of the wings and their contraction elevates or depresses the wing, altering its angle during flight.
- In smaller-winged insects the action is indirect: muscles move the thoracic plates which in turn move the wings, allowing very rapid wing-beating.

HYDROSTATIC SKELETONS

Many soft-bodied animals are supported by fluid contained in the body cavity. The fluid exerts pressure on the body muscles, which in turn are able to contract against the fluid, bringing about locomotion and maintaining the animal's shape.

EARTHWORMS

This system is exemplified by earthworms. In these animals the body segments are separated by septa and so pressure is localised and segments can expand or contract independently. There are two antagonistic sets of muscles, outer **circular** and inner **longitudinal**, which bring about extension and contraction respectively. By co-ordination of the activity of these muscles and the protraction or retraction of the **chaetae**, an earthworm can move either forwards or backwards. Segments extend and contract in sequence. Anterior segments elongate, the chaetae are protracted to grip the soil and at the same time the posterior chaetae are retracted and the segments contract. This process is repeated so that the worm moves by waves of alternate muscle contractions. The contractions are controlled by nerve impulses. Rapid withdrawal movements can be produced by stimulation of epidermal sense organs at the anterior and posterior ends of the animal. The impulses for these protective movements are transmitted by special large nerve fibres running the length of the ventral nerve cord.

AMOEBOID MOVEMENT

This takes place in a variety of non-cellular animals, mainly in the class Rhizopoda and also in various types of motile cells in multicellular animals. Amoeboid cells move by forming temporary projections, **pseudopodia**, into which the rest of the cell flows.

The fountain-zone theory

The most favoured theory for amoeboid movement is the **fountain-zone** theory. At the point where a pseudopodium develops, the ectoplasm changes from gel to sol and flows backwards. At the same time the endoplasmic sol flows forward and changes to gel. By this alternation of changes from sol-gel-sol the cell moves forwards by cytoplasmic streaming. It is thought that the protein molecules (actin filaments) which are extended in the plasma sol contract, when the sol changes to gel, and so pull the cell forward. These changes are shown in Fig. 14.10.

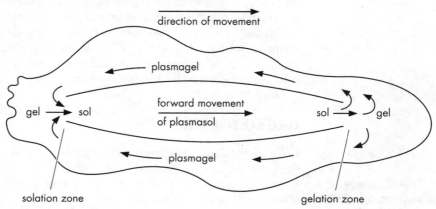

Fig. 14.10 Fountain-zone theory

CILIA AND FLAGELLA

These structures have a similar structure of two central filaments surrounded by nine pairs of filaments round the periphery. The peripheral filaments are made of the protein **tubulin**. One filament of each pair has two arms made of another protein, **dynein**, which is an ATPase and releases the energy for the contraction of the organelle. This structure is shown in Fig. 14.11.

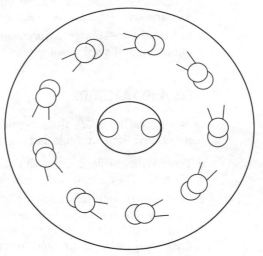

Fig. 14.11 Structure of a cilium

CILIA

Cilia are usually found in groups or over the whole surface of a cell. When active they beat asymmetrically; there is a fast down-beat with the cilium straight, then a slow recovery stroke with the cilium limp. Their activity is co-ordinated so that they beat in waves along the length of the cell. This phenomenon is known as **metachronal rhythm**.

FLAGELLA

Flagella are usually found singly or in pairs. Undulations occur along the length of the flagellum, driving the organism forward. A flagellum may be at the front of the cell, thus pulling it along, or at the rear producing a push. In some organisms, such as *Euglena*, the flagellum spins the cell as well as moving it forward, so that the overall movement is along a helical path.

> Be sure you can distinguish between cilia and flagella. The terms are not

SUPPORT TISSUES IN PLANTS
synonyms

Plant organs are supported by a variety of different tissues the occurrence and distribution of which varies both with the organ concerned and with whether the plant is herbaceous or woody.

PARENCHYMA CELLS

Parenchyma cells are living cells with thin cellulose walls. They carry out a variety of functions, one of which is to give support by means of their turgidity. When fully turgid the cells press against each other and provide mechanical support to the plant organ in which they are found. Leaves are largely supported by parenchyma cells, as are the stems of herbaceous plants. When a leaf or herbaceous stem is short of water the parenchyma cells lose turgor and the plant is seen to wilt. Normal support is established when water is once again available.

COLLENCHYMA CELLS

> Remember that collenchyma cells are not lignified

Collenchyma cells are living cells with cellulose walls which have extra thickening at the corners. They are found in the outer cortex of many stems and in the midribs of leaves. Columns of collenchyma are found in the angles of square stems, forming strengthening girders for the stem.

SCLERENCHYMA CELLS

Sclerenchyma cells are dead cells with very thick lignified walls. There are two forms of them.

- **Fibres** are elongated cells with tapering ends which interlock. They may form part of the xylem and phloem but are also found in the outer cortex of stems or forming caps or sheaths to the vascular bundles in herbaceous stems.
- **Sclereids** are similar to fibres and are found scattered in some stems, leaves, fruit endocarp and seed coats.

VESSELS AND TRACHEIDS

Vessels and tracheids are dead, lignified cells found in the xylem. They give support because of their lignified walls, but are mainly involved in the transport of water and salts.

- **Tracheids** are single cells with tapering end walls which overlap, giving mechanical strength. They are the only cells found in the xylem of the primitive vascular plants.
- **Vessels** are long tubular structures formed by the fusion of cells end to end. The end walls of the fused cells break down so that there is an unobstructed lumen through the whole length of the vessel. In **protoxylem** vessels the walls are thickened with lignin laid down in an annular or spiral pattern, whilst in **metaxylem** the thickening is heavier, being in either a reticulate pattern or complete except for the pits between adjacent vessels.

XYLEM DISTRIBUTION

The distribution of xylem tissue differs in primary stems, leaves and roots.

- In **stems** it occurs as part of the peripheral vascular bundles. This organisation gives flexible support, but also resistance to bending strain.
- In **leaves** the arrangement of vascular tissues in the midrib and network of veins also gives flexible strength and resistance to shearing strains.
- In **roots** the central arrangement is ideal for resistance to pull and so helps the anchorage of the plant.

SECONDARY XYLEM

When the plant stems and roots become woody, the amount of xylem tissue is increased during the process of secondary thickening. Secondary xylem is laid down in concentric rings by the cambium forming a supporting cylinder for the growing plant.

The different arrangements of xylem tissue are shown in Fig. 14.12.

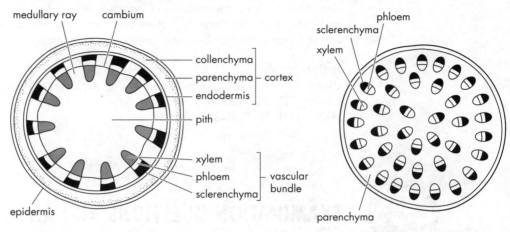

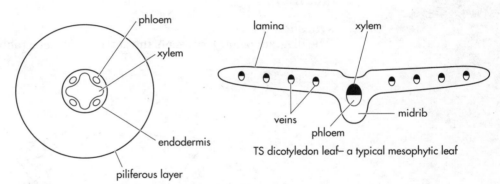

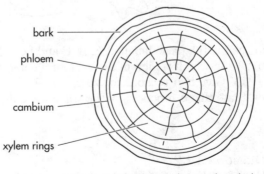

Fig. 14.12 Arrangements of xylem in different plant structures

PRACTICAL WORK

Most syllabuses include the histology of bone, muscle and sometimes cartilage. You should study these tissues under both the low and high power of the microscope. This work should include a comparison of the structure of smooth, striated and cardiac muscle. The histology of flowering-plant support tissues should also be revised. You should study transverse and longitudinal sections of plant stems, transverse sections of leaves and roots and also macerated plant tissue.

A detailed study of vertebrae, girdles and the main limb bones of a mammal is necessary as questions may include detail of a single bone or a comparison of bones. This would be either in a theory paper or as part of a practical assessment.

A study of at least one set of antagonistic muscles, usually the biceps and triceps, is necessary. You should know the origins and the insertions of the muscles. If your syllabus includes specific examples of animal movement, you should know the muscles involved and how they work together to produce the movement. Videotapes and films are useful in this context.

A microscopic study of living amoeba, a ciliate and a flagellate is necessary to understand these movements. Observations of live earthworms will help you to understand the relationship between the circular and longitudinal muscles, chaetae and the compartmentalised hydroskeleton in bringing about locomotion.

Further reading

Clegg C J and Cox H, 1978, *Anatomy and Activities of Plants*. John Murray
Currey J D, 1970, *Animal Skeletons*. Arnold
Wilkie D R, 1976, *Muscle*. Arnold

EXAMINATION QUESTIONS AND ANSWERS

QUESTIONS

Question 14.1
The diagram below represents the anterior view of a rabbit vertebra.

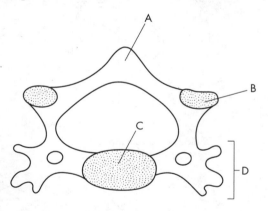

Fig. 14.13

a) Name the parts labelled A, B, C and D. (4)
b) i) In which region of the vertebral column is this vertebra found? (1)
   ii) Give **two** reasons for your answer. (2)
c) State **three** functions of the vertebral column in a terrestrial mammal. (3)

(Total 10 marks)
(ULEAC)

Question 14.2
Here is a diagram of the joint between the femur and the pelvic girdle of a mammal.
a) State the functions of the following structures in allowing movement of the femur.
   i) Cartilage
   ii) Synovial fluid
   iii) Ligament (6)
b) Explain how the arrangement of muscles can bring about movement of the femur in one plane only. (Names of individual muscles are not required.) (6)

(Total 12 marks)
(UCLES)

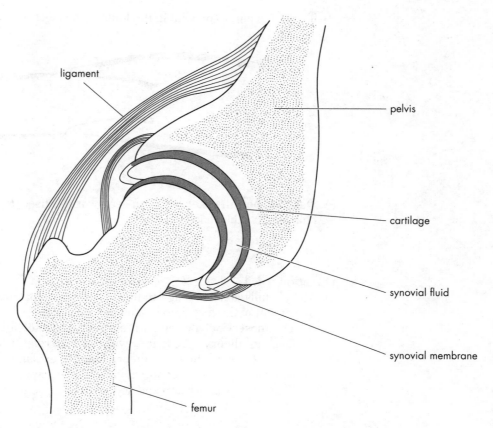

Fig. 14.14

Question 14.3
Muscles achieve movement by reacting against each other and against a skeleton.
a) What type of skeleton occurs in:
   i) a mammal
   ii) an earthworm
   iii) an insect? *(3)*
b) In the table below state
   i) where in the limb of a mammal you would expect to find tendons and ligaments
   ii) whether or not the tendon and ligament are elastic.

	(I) SITE	(II) ELASTICITY
Tendons		
Ligaments		

c) What properties of muscle tissue require that muscles work in antagonistic pairs? *(2)*
d) On the diagram below, draw in the muscles and their attachments that cause the following movements
   i) the movement of the human arm at the elbow (label the muscles you have drawn)

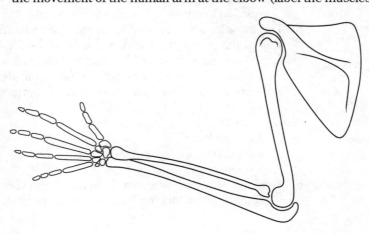

Fig. 14.15

CHAPTER 14 SUPPORT, MOVEMENT AND LOCOMOTION

    ii) the flexing of the joint in the limb of an insect   (6)

Fig. 14.16

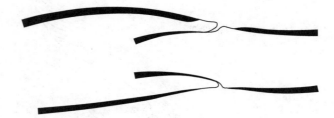

e) Describe how the skeleton of an earthworm is used, together with the muscles and chaetae, to achieve locomotion.   (4)

(Total 15 marks)
(ULEAC)

Question 14.4
a) Make a fully labelled diagram to show the principal muscles and bones involved in movement of the fore limb of a named mammal.   *(8)*
b) i) The most widely accepted theory of muscle contraction is termed 'the sliding filament theory'. Describe, with special reference to the micro-anatomy of striated muscle, how this theory helps to explain muscle contraction.   *(16)*
   ii) In order to initiate contraction, a nerve impulse must arrive at the muscle. Explain how potassium and sodium ions are important in this process.   *(6)*

(Total 30 marks)
(ODLE)

OUTLINE ANSWERS

Answer 14.1
a) A = neural spine
   B = anterior zygapophysis
   C = centrum
   D = transverse process
b) i) Cervical/neck region
   ii) Very short neural arch; presence of vertebrarterial canals; neural canal large compared with the centrum
c) Support of the body mass; attachment of muscles; protection of spinal cord

Answer 14.2
a) i) Cartilage covers the ends of the bones. It can withstand weight and absorb mechanical shocks. It is smooth and helps to reduce friction.
   ii) Synovial fluid acts as a lubricant for the joint surfaces, reducing friction between them.
   iii) A ligament joins a bone to a bone. It is elastic and so allows one bone to move in relation to another. The ligaments together form a fibrous capsule for the joint.
b) The muscles are arranged in antagonistic sets. A muscle is attached to the dorsal side of the pelvis and the dorsal side of the femur shaft. When it contracts it will move the femur dorsally. The antagonistic muscle is attached to the ventral side of the pelvis and to the ventral side of the femur shaft. When this contracts the femur will swing ventrally. When either muscle contracts, its antagonist relaxes.

Answer 14.3
a) i) Endoskeleton
   ii) Hydrostatic skeleton
   iii) Exoskeleton
b) Tendons: site attaching muscle to bone; non-elastic
   Ligaments: site attaching bone to bone; elastic

c) Muscles exert a force when contracting; no force is exerted, muscle is non-functional when relaxing.

d) i)

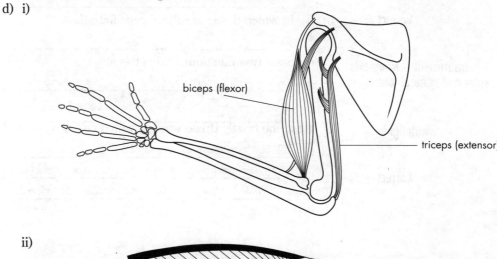

Fig. 14.17

ii)

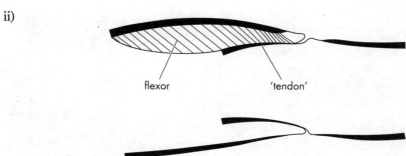

Fig. 14.18

e) Skeleton is incompressible; circular muscles contract and chaetae are retracted; segments elongated; chaetae protracted and longitudinal muscles contract, pulling the worm forwards; this process repeats in sequence down the body.

QUESTION, STUDENT ANSWER AND EXAMINER COMMENTS

Question 14.5
The diagrams below show cells from a tissue of a flowering plant.

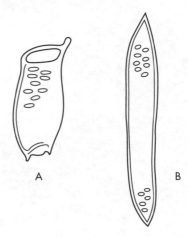

Fig 14.19

a) i) Name cells A and B.

A Vessel

B Tracheid   66 Correct 99   (2)

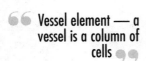

66 Vessel element — a vessel is a column of cells 99

ii) State **one** way in which the structure of cell A differs from cell B.

Correct → It is open at both ends (1)

b) i) In which tissue are these cells found?

Xylem (1)

ii) State **two** functions of this tissue.

Insufficient – transports water and mineral ions → Transports water

Gives support *Correct* (2)

c) Describe briefly **three** ways in which cell A is adapted for its functions in the plant.

... with lignin → Its wall is thickened to give support.

The cell contents have been lost giving a hollow lumen.

Correct → End walls broken down so water flow is not interrupted. (3)

(Total 9 marks)
(ULEAC)

CHAPTER 14 **REVIEW SHEET** 199

REVIEW SHEET

1 Complete the labels for each of the following diagrams.

pectoral girdle   **pelvic girdle**

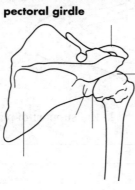 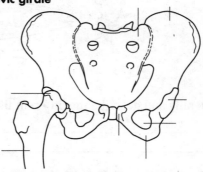

2 Write briefly on the *functions* of each of the following.
   a) Pectoral girdle: _____
   _____
   b) Clavicle: _____
   _____
   c) Pelvic girdle: _____
   _____

3 Describe the structure and function of each of these joints (giving appropriate examples).
   a) Immovable joints: _____
   _____
   b) Partially movable joints: _____
   _____
   c) Synovial joints: _____
   _____

4 Briefly outline the process by which **bone** is laid down.
   _____
   _____
   _____

5 Explain the role of the two *proteins* tropomyosin and troponin in the **contraction of muscles**.
   _____
   _____
   _____

6 Write briefly on each of the following.
   a) Walking in insects: _____
   _____
   b) Flight in insects: _____
   _____

7 Complete the labels on the following diagram (include appropriate labelling of arrows). Then use the diagram to explain the **fountain-zone theory** of amoeboid movement.

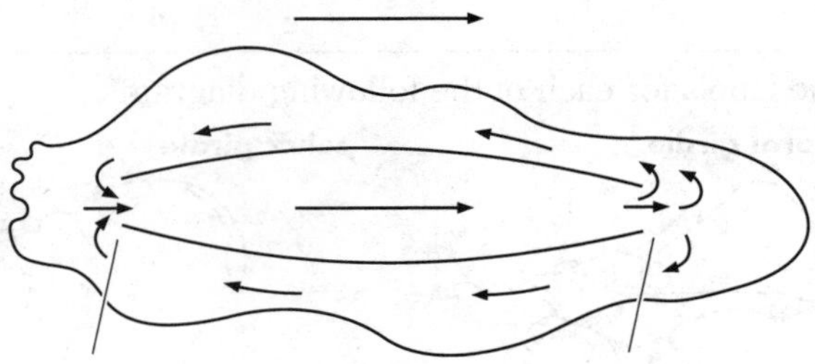

8 Identify and describe the different **tissues** involved in the support of plant organs.
_____
_____
_____
_____
_____

9 Use the **earthworm** as an example to illustrate the locomotion of soft-bodied animals.
_____
_____
_____

10 Describe the different possible arrangements of **xylem tissue** in the following. Draw a diagram to illustrate your answer in each case.

a) Stems: _____
_____
_____
_____
_____
_____

b) Leaves: _____
_____
_____
_____
_____
_____

c) Roots: _____
_____
_____
_____
_____
_____

CHAPTER 15

GENETICS AND EVOLUTION

- VARIATION
- MITOSIS AND MEIOSIS
- GENE CONCEPT
- MENDEL'S LAWS
- CHROMOSOMAL THEORY OF INHERITANCE
- POLYPLOIDY
- LINKAGE AND CHROMOSOME MAPPING
- SEX DETERMINATION AND LINKAGE
- MUTATION
- POPULATION GENETICS
- ORGANIC EVOLUTION
- SELECTION
- ORIGIN OF SPECIES
- CONSERVATION OF GENE POOLS
- PRACTICAL WORK

GETTING STARTED

The topic of genetics is concerned with the processes by which the biochemical basis of an organism's characteristics is handed on from generation to generation. This information, which is principally, for the production of polypeptides, is coded in the **genes**. Before continuing with this chapter it would be wise to refer back to Chapter 4, on Cell Biology, to revise the process of genetic coding and the translation of the code into the polypeptides which form the proteins in the cells.

This chapter is concerned with how genes behave in organisms, how they interact with each other and how their effects may be modified by environmental influences. These are the phenomena which bring about variation between individuals of a species. This variation is the raw material of evolution, and the interconnection between genetics and evolution is considered. Many questions involve distinguishing between different genetic phenomena or solving genetics problems so practice in solving problems (setting out the steps in a logical sequence) is advisable. The part played by variation in the process of evolution plus the presence of recessive genes in gene pools is also an important aspect of your A-level work.

CHAPTER 15 GENETICS AND EVOLUTION

ESSENTIAL PRINCIPLES

VARIATION

This term is used to describe the difference in characteristics of members of the same species.

DISCONTINUOUS VARIATION

> You must not confuse the terms gene and allele

Discontinuous variation is usually controlled by one or two genes of major effect, each having two or more **allelic** forms. For example, the light and dark forms in some moth species or the four alternative ABO blood groups in human beings. Organisms show two or more discrete alternatives in some characteristics.

CONTINUOUS VARIATION

Other characteristics of species show continuous variation, which is a continuum of minor differences from one end of a range to the other. This continuous variation is shown by characters such as shape or linear dimensions and is produced by the influence of many genes, each having a small effect, and also by environmental factors. For example, an organism will inherit genes, giving it a theoretical maximum size, but whether or not this is reached will depend upon nutrition during the growth period and other environmental factors.

The differences between the members of a species are produced by a combination of continuous and discontinuous variations. Different members of the species will have different combinations of alleles and will have been subject to different environmental influences.

MITOSIS AND MEIOSIS

There are two forms of nuclear division.

- **Mitosis** is concerned with producing new cells for growth and repair or with reproduction in non-cellular or unicellular organisms which undergo binary fission.
- **Meiosis** is the reduction division in which pairs of chromosomes in diploid organisms are separated in the formation of **haploid** gametes.

Mitosis is not directly concerned in genetic phenomena but examination questions based on cell division often involve a comparison of the two types of division.

MITOSIS

> Be very careful with spelling or mitosis and meiosis will be confused

This is a gradual process but is classically subdivided into four descriptive stages with an interphase between successive periods of division (see Fig. 15.1).

Interphase
Interphase is the longest part of the cycle during which a newly formed cell increases in size and produces organelles lost during the previous division. Just before the next cell division the chromosomes replicate so that each then consists of two **chromatids** joined together by the **centromere**.

Prophase
Prophase is the phase of mitotic division during which the chromatids shorten and become thicker by the spiralisation and condensation of the DNA protein coat. In the cells of animals and lower plants, the **centrioles** move to the poles of the cells and microtubules begin to radiate from them, forming **asters**. At the end of prophase the nuclear membrane disintegrates and the **spindle** is formed. During this phase the nucleoli become smaller.

Metaphase
In metaphase, the pairs of chromatids attach to the spindle by microtubules, **spindle fibres**, at their centromeres and move along the spindle until they line up along the equator of the cell at right angles to the spindle axis.

Anaphase
The anaphase stage is very rapid. The centromeres split and the spindle fibres pull the now separated chromatids to the poles of the cell, where they become the chromosomes of the two daughter cells.

Telophase
Mitosis ends with telophase. The chromosomes, having reached the poles of the cells, uncoil and lengthen. The spindle breaks down, the centrioles replicate, the nucleoli reappear and the nuclear membrane reforms. In plant cells, a cell plate forms across the middle of the cell and a new cell wall is laid down.

Function of mitosis
Mitosis produces two cells which have the same number of chromosomes as the parent cell and each chromosome is an exact replica of one of the originals. The division, therefore, allows the production of cells which are genetically identical to the parent and so gives genetic stability. By producing new cells, mitosis leads to growth of an organism and also allows for repair of tissues and the replacement of dead dells. Note that in higher plants only meristematic cells normally carry out mitosis, whilst in animals the majority of cells can divide and produce new cells if the need arises. The other function of mitosis is to provide for asexual reproduction.

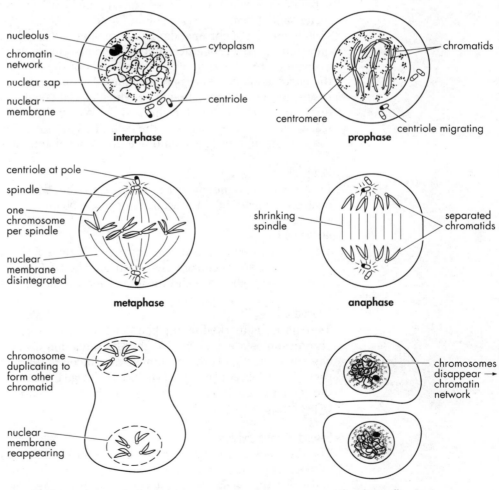

Fig. 15.1 Stages of mitosis

MEIOSIS

This is the reduction division which occurs during gamete formation or in the formation of haploid spores in sexually reproducing organisms. In this division the diploid number of chromosomes (2n) is reduced to the haploid (n). Usually there are two cycles of division:

- the first is where the chromosome number reduces
- the second is where the two new haploid nuclei divide again in a division identical to that of mitosis
- the result is that four haploid nuclei are formed from the parent nucleus

In males
In males of both animals and plants all four products of meiosis survive, forming spermatozoa and pollen grains respectively.

In females
Meiosis in females gives rise to just one egg cell from the four products.

Like mitosis, meiosis is a gradual process but for description it is divided into the four phases of prophase, metaphase, anaphase and telophase, these phases occurring once in each of the two divisions.

Prophase I
Prophase I is the longest phase and is now described as a continuum instead of five separate phases. Chromosomes pair up, the identical (**homologous**) chromosomes of maternal and pattern origin coming together. At this stage each chromosome can be seen to consist of chromatids. The homologous chromosomes now seem to repel each other but are held together at certain points known as **chiasmata**. These are sites of exchange of material between chromatids of the homologous pair, brought about by the chromatids breaking and rejoining. This process is known as **crossing over** and results in the recombination of genetic material (see Linkage). Note that crossing over does not occur in all sister chromatids. Whilst these chromosome events are taking place, other changes occur in the cell. The centrioles, if present, migrate to the poles, the nucleoli and nuclear membrane break down and the spindle forms.

Metaphase I
In metaphase I the bivalents migrate to the equator and come to lie with their centromeres, oriented towards the opposite poles.

Anaphase I
When anaphase I begins the centromeres of each bivalent do not divide but the spindle fibres pull whole centromeres, each attached to two chromatids, towards opposite poles, so separating the chromosomes into two haploid sets.

Telophase I
Telophase I is marked by the pairs of chromatids reaching the poles, half the original chromosomes are at each pole and the reduction division is complete. In animal cells and some plant cells the chromosomes uncoil and a nuclear membrane is formed at each pole. Cleavage in animals cells or cell wall formation in plant cells then occurs and the cells may go into interphase. Most plant cells however go straight from anaphase I to prophase II.

Second meiotic division
The mechanical processes of the second meiotic division are very similar to those of mitosis. The second division results in the sister chromatids separating in anaphase II and when the nuclei reform in telophase II the net result is four nuclei, each with a haploid number of chromosomes.

Features of meiosis
Where crossing over has occurred at the chiasmata formed in prophase I, the four chromosomes formed from each original pair are all genetically different. This is one

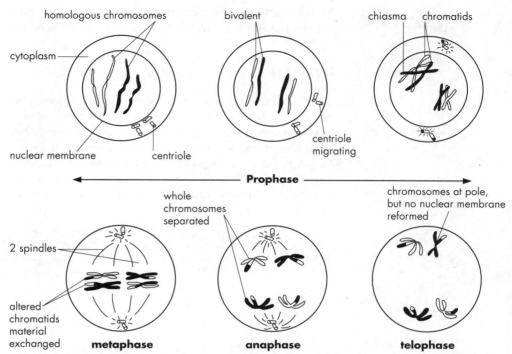

Fig. 15.2 Stages of meiosis I

of the major causes of genetic variation. It is one of the significant features of meiosis and is an important different from mitosis, which produces genetically identical cells. The other significance is the production of haploid not diploid gametes, thus ensuring the maintenance of the diploid condition when gametes fuse at fertilisation.

GENE CONCEPT

In 1866 Gregor Mendel suggested that the characteristics of organisms were determined by units or 'elements' which were handed on from generation to generation. Later these units were identified as **genes** which were carried on and transmitted by chromosomes.

DEFINITION OF A GENE

Although we know that genes consist of DNA and the structure of DNA is known, it is still difficult to give a precise definition of a gene. Genes have three main characteristics:

- they can separate and recombine
- they can mutate
- they code for the production of specific polypeptides

Any of these can be used to define a gene but they all have limitations.

- If a gene is defined as the smallest segment of a chromosome which can separate and recombine, it is a relatively large unit which may contain information for making more than one polypeptide.
- Defined as the smallest segment which can mutate, a gene would be a single base pair, which would be only a tiny fraction of a polypeptide code.

The **definition of a gene** based on function has been refined to *one gene being the portion of a chromosome which codes for one polypeptide*. In present terminology this piece of code for one polypeptide is called a **cistron**, which consists of a sequence of base pairs, **mutons**, these being the units which can mutate. Three mutons form a **codon** for one amino acid and so a cistron is formed of a series of codons, each formed of three mutons.

Note that even this definition does not tell the whole story as some genes do not give rise to a product but control the function of other genes. Particular genes are found at specific sites on chromosomes, each of which is a **locus**. The alternative forms of a gene which may be found in the corresponding position on homologous chromosomes are known as **alleles**.

MENDEL'S LAWS

Mendel started the first documented scientific investigation of inheritance and formulated two laws, which are basic to the science of genetics.

FIRST LAW

Mendel's early experiments were based on selecting pea plants of two varieties which showed clearly separable characteristics, such as tall and dwarf plants, round and wrinkled seeds and axial and terminal flowers. From his results he formulated the **law of segregation**, which states that *'the characteristics of an organism are determined by factors which occur in pairs. Only one member of a pair of factors can be represented in any gamete'*. The inheritance of a single pair of contrasted characters is known as **monohybrid inheritance**. The process can be represented in a genetic diagram (see Fig. 15.3).

T represents the allele for tall plants (dominant)
t represents the allele for dwarf plants (recessive)
$2n$ is the diploid state with pairs of chromosomes
n is the haploid state with single chromosomes

Parental phenotypes	Pure breeding tall plants		Pure breeding dwarf plants	
Parental genotypes ($2n$)	TT		×	tt	
Gametes	T	T	t	t
F_1 genotype ($2n$) by random fertilisation	Tt	Tt	Tt	Tt
	F_1 genotypes all heterozygous F_1 phenotypes all tall			
Self-pollination of the F_1 plants	Tall plant		Tall plant	
F_1 genotypes ($2n$)	Tt		×	Tt	
Gametes (n)	T	t	T	t
F_2 genotypes ($2n$) by random fertilisation	TT	Tt	Tt	tt
	Homozygous	Heterozygous		Homozygous
	F_2 phenotypes 3 tall: 1 dwarf			

Fig. 15.3 Genetic diagram: monohybrid cross

Dominant allele
Of the pair of alleles in a monohybrid cross, the one that always produces an effect on the appearance of the organism when present is the dominant allele. It is usually represented by a capital letter.

Recessive allele
The allele that produces an effect only when present as an identical pair is the recessive allele, represented by a lower-case letter. For example, in Fig 15.3 the allele for the tall characteristic which is dominant is represented by T and that for the dwarf characteristic by t.

Genetic crosses – terms
There are specific terms used in connection with genetic crosses. The combination of alleles found in an individual is the **genotype**. If both alleles are the same, the genotype is described as **homozygous**, whilst if the pair is dissimilar it is known as **heterozygous**. The appearance of an organism, determined by the genotype, is the **phenotype**. Because a dominant allele always produces a phenotypic effect if present, it follows that one phenotype may be produced by more than one genotype.

Example: monohybrid cross
In Fig. 15.3 both TT and Tt give the tall phenotype. It follows from this that in a monohybrid cross, three different genotypes but only two phenotypes are produced. Because of random pairing of gametes at fertilisation, two heterozygotes are formed along with one homozygous dominant and one homozygous recessive genotype in the second (F_2) generation. The overall phenotypic ratio is 3 dominant to 1 recessive. This generation is produced by crossing (or in Mendel's case self-pollinating) the F_1 hybrids produced from the pure-breeding parents.

SECOND LAW

Having worked out the method of inheritance of a single pair of contrasted characters, Mendel investigated what happened if he used plants which differed by two pairs.

Experiments

In one of his experiments, Mendel worked with the **colour** and **shape** of the **pea** cotyledons. He crossed homozygous (pure-breeding) plants with round and yellow cotyledons in their seeds with homozygous plants with wrinkled and green cotyledons. When plants grown from these seeds were self-pollinated, the seeds produced were of four different types of shape and colour of cotyledon. Altogether he collected 556 seeds. The four types were:

- round yellow
- wrinkled yellow
- round green
- wrinkled green

The other point which interested him was that they were found in the ratio of approximately 9:3:3:1, proportions which are now known as the **dihybrid ratio**. He knew from his previous experiments that round was dominant over wrinkled and that yellow was dominant over green. These results showed that each pair of characters still gave a 3:1 ratio, as in monohybrid inheritance, but also that they recombined independently. This led him to formulate his **second law**, stating the *'either one of a pair of characteristics may combine with either of another pair'*. The probability of the four alleles appearing in any of the F_2 offspring is:

- round (dominant) $\frac{3}{4}$
- wrinkled (recessive) $\frac{1}{4}$
- yellow (dominant) $\frac{3}{4}$
- green (recessive) $\frac{1}{4}$

The probability of combinations of alleles appearing in the F_2 is therefore as follows:

round and yellow = $\frac{3}{4} \times \frac{3}{4} = \frac{9}{16}$
round and green = $\frac{3}{4} \times \frac{1}{4} = \frac{3}{16}$
wrinkled and yellow = $\frac{1}{4} \times \frac{3}{4} = \frac{3}{16}$
wrinkled and green = $\frac{1}{4} \times \frac{1}{4} = \frac{1}{16}$

Fig. 15.4 is a genetic diagram to explain the results which Mendel obtained, represented in a **Punnett square**. Using this square minimises the risk of making mistakes, especially if you fill in all the male gametes and then add the female, or vice versa.

R represents round seed (dominant)
Y represents yellow seed (dominant)
r represents wrinkled seed (recessive)
y represents green seed (recessive)

Parental phenotypes	Pure breeding round yellow	Pure breeding wrinkled green
Parental genotypes (2n)	RRYY	rryy
Gametes (n)	all RY	all ry
F_1 genotype (2n) random fertilisation	all RrYy	
Self fertilisation	RrYy × RrYy	
Gametes identical from ovules and pollen grains	RY Ry rY Yy	

F_2 genotypes (2n) and phenotypes shown in a Punnett square

9 round yellow
3 round green
3 wrinkled yellow
1 wrinkled green

♀ \ ♂	RY	Ry	rY	ry
RY	RY RY round yellow	Ry RY round yellow	rY RY round yellow	ry RY round yellow
Ry	RY Ry round yellow	Ry Ry round green	rY Ry round yellow	ry Ry round green
rY	RY rY round yellow	Ry rY round yellow	rY rY wrinkled yellow	ry rY wrinkled yellow
ry	RY ry round yellow	Ry ry round green	rY ry wrinkled yellow	ry ry wrinkled green

Fig. 15.4 Genetic diagram: dihybrid cross

DEVIATIONS FROM MENDEL'S LAWS

Not all the characteristics are controlled by single genes which behave independently, as was the case in Mendel's experiments. The following subsections describe some of the gene phenomena which show deviations from Mendel's laws.

Incomplete dominance (co-dominance)

In many organisms there are characteristics controlled by pairs of alleles where one allele is not dominant over the other but in combination both have phenotypic effect. This is known as incomplete dominance, co-dominance or blending. In most cases, the heterozygote shows a phenotype intermediate between those of the two homozygotes. Examples of incomplete dominance are:

- antirrhinums – red (RR), pink (Rr) and white (rr) flowers
- shorthorn cattle – red (RR), roan (Rr) and white (rr) coat colour

The genetic diagram for these crosses is the same as that illustrating Mendel's first law (see Fig. 15.3) but in the F_1 all individuals have the intermediate phenotype, as do the heterozygotes in the F_2. So, for example, in a cross between pure-breeding red and pure-breeding white antirrhinums, in the F_2; red, pink and white flowered plants are found in the ratio of 1:2:1.

Multiple alleles

Many characteristics in organisms are controlled not by pairs of alleles but by three or more alleles of which any two may occupy the gene loci on the homologous chromosomes. Examples of sets of phenotypes controlled by multiple alleles are:

- coat colour in rodents
- the ABO blood groups in humans

Blood groups are controlled by three alleles of a gene which govern the formation of red blood cell protein, **isohaemagglutinogen**. The two alleles I^A and I^B are co-dominant and the allele i is recessive. The possible genotypes and corresponding phenotypes are shown in Table 15.1.

GENOTYPE	PHENOTYPE (BLOOD GROUP)
$I^A I^A$	A
$I^A i$	A
$I^A I^B$	AB
$I^B I^B$	B
$I^B i$	B
ii	O

Table 15.1 Human ABO blood groups

Polygenes

Some characteristics are controlled by several pairs or sets of alleles found at different loci, each having a small effect on the phenotype. This is polygenic (or multiple gene) inheritance. Many human characteristics are controlled by polygenes: for example, height, skin colour and intelligence. Note that often there are also marked environmental influences on these characters, giving overall continuous variation in a population.

> 66 Do not confuse polygenes, which are at different loci, with multiple alleles, all members of a set belonging to a single locus 99

Gene interaction

With gene interaction a phenotype is produced by two or more pairs of alleles acting together, an example being comb shape in domestic hens.

- Domestic hen comb shape: there are four different comb shapes, known as rose, pea, walnut and single. Rose comb is produced by the dominant gene R and pea by the dominant gene P. If both R and P are present, walnut comb is produced and if both are absent there is a single comb. Thus there are two pairs of genes involved in producing comb shape. The genetics of domestic hen comb shape is explained in Fig. 15.5.

> 66 Remember that this involves sets of alleles from different loci 99

R represents rose comb (dominant)
r represents single comb (recessive)
P represents pea comb (dominant)
p represents single comb (recessive)

Parental phenotypes	Pure breeding rose comb		Pure breeding pea comb
Parental genotypes (2n)	RRpp	×	rrPP
Gametes (n)	all Rp		all rP
F_1 genotype (2n)		all RrPp	
random fertilisation			
F_1 phenotype (2n)		all walnut comb	
cross fertilisation F_1		RrPp × RrPp	

Gametes identical for eggs and sperm

F_2 genotypes (2n) and phenotypes shown in a Punnett square

9 walnut
3 rose
3 pea
1 single

♀ \ ♂	RP	Rp	rP	rp
RP	RP RP walnut	Rp RP walnut	rP RP walnut	rp RP walnut
Rp	RP Rp walnut	Rp Rp rose	rP Rp walnut	rp Rp rose
rP	RP rP walnut	Rp rP walnut	rP rP pea	rp rP pea
rp	RP rp walnut	Rp rp rose	rP rp pea	rp rp single

Fig. 15.5 Genetic diagram: polygenic inheritance of comb shape in domestic hens

CHROMOSOMAL THEORY OF INHERITANCE

The basis of this theory is that pairs of alleles are carried on pairs of homologous chromosomes. Mendel's laws are explained by the separation and recombination of pairs of alleles. These phenomena can in turn be explained by the movement of chromosomes during meiosis. The random alignment of homologous chromosomes at metaphase I of meiosis and their separation as the division proceeds gives rise to the different allele combinations in the resultant gametes. The separation of homologous chromosomes explains segregation and the randomness forms the basis of independent assortment.

- In *Drosophila*, which has four pairs of chromosomes, the number of different possible chromosome combinations in gametes is 2^4 (16).
- In *Homo*, with 23 pairs, there are 2^{23} possible combinations, which is over 8 million.

Further evidence for the chromosome theory is furnished by experimental results and genetic observations which show deviations from Mendel's laws. In each case there is a chromosomal phenomenon which ties in with this observation (see later sections, Linkage and Chromosome Mutations).

POLYPLOIDY

This is a term used to describe cells which contain three or more sets of chromosomes: 3N is **triploid**, 4N is **tetraploid**, etc. The phenomenon is much more common in plants than in animals. The extra sets of chromosomes interfere with gamete formation, and therefore polyploids often have low fertility, but many plants can reproduce vegetatively, so the lowered fertility is less important. Polyploidy often gives increased size, vigour and disease resistance – about half the known species of Angiosperms show evidence of an origin involving polyploidy. There are two different sorts of polyploidy.

AUTOPOLYPLOIDY

In this form there is an increase in the number of chromosome sets within the species. For example, if nuclear division takes place but the cytoplasm fails to cleave, a tetraploid cell results. This cell will divide to form two more tetraploid cells. These cells are larger in size as the amount of cytoplasm increases to preserve the nucleus/cytoplasm volume ratio and the result is that the whole plant pincreases in size. Autopolyploids may arise naturally but they can be produced artificially by using drugs, such as colchicine, which prevents spindle formation.

- Many important **crop plants** are autopolyploids: for example, bananas and sugar beet.

- Because of the lack of fertility, **animal** autopolyploids are not known but sometimes polyploid tissues are found within an animal. An example is the tetraploid cells in the human liver.

ALLOPOLYPLOIDY

In this form two related species form a sterile hybrid whose chromosomes become doubled to form a fertile tetraploid, trebled to form a hexaploid, etc. These polyploid species are fertile if crossed together but cannot reproduce with either of the original parents. Most allopolyploids have chromosome numbers which are multiples of the diploid numbers of the parental species.

An example of allopolyploidy is found in the salt marsh grass genus *Spartina*. *S maritima* (2n = 60) and *S. alterniflora* (2n = 62) formed a hybrid, *S. townsendii* (2n = 61). This grass was sterile and reproduced by rhizomes. Some time in the past 50 years a fertile allopolyploid, *S. anglica* (2n = 122), arose and is replacing the parental species.

> **Chromosome doubling must occur to form a fertile allopolyploid**

Most allopolyploids are phenotypically different from the parents and many are important **crop plants**: for example, **wheat**. By crossing with other wild grasses, the original 'einkorn' wheat (2n = 14) has given rise to various, more productive, crop wheats: for example, *Triticum durum* (2n = 28), which is a pasta wheat, and *T. aestivum* (2n = 42), which is bread wheat. The derivation of *T. aestivum* is shown in Fig. 15.6. Other examples are plums, loganberries and oats.

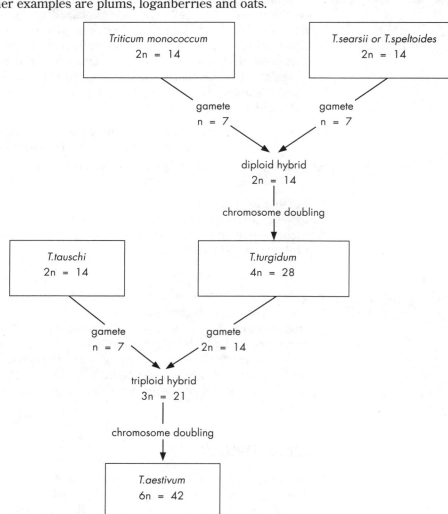

Fig. 15.6 Origins of modern bread wheat

LINKAGE AND CHROMOSOME MAPPING

LINKAGE GROUPS

When the loci for two genes are found on the same chromosome. these genes are said to be linked. All the genes on a single chromosome are linked together and form a linkage group. They are on the same chromosome, so usually they pass into the same gamete and are inherited together.

CROSSING OVER

If we consider two pairs of alleles carried on a pair of homologous chromosomes and producing contrasting characters, instead of the Mendelian ratio of 9:3:3:1 in the F_2 generation, they tend to produce either a 3:1 or 1:2:1 ratio as the two alleles on each chromosome usually stay together. However, total linkage is rare and usually in breeding experiments there is some '**unlinking**'. When chiasmata are formed in the prophase of meiosis I and exchange takes place between a pair of homologous chromosomes, members of the allelic pairs are recombined, producing a small number of new phenotypes unlike either parent. This is the result of crossing over. When a breeding experiment involving two pairs of alleles gives an unusual ratio in the F_2 generation, linkage is suspected and this can be checked by carrying out a **test-cross** experiment.

Example: test-cross

Fig. 15.7 shows a test-cross between a *Drosophila* heterozygous for body colour and wing length and one showing the homozygous recessive phenotype. The resulting phenotypic ratio indicates that the two pairs of alleles are linked and that the cross-over value between them is 17%.

Test-cross phenotypes	Grey body long wing heterozygote		Ebony body vestigial wing homozygote	
Test-cross genotypes (2*n*)	$\frac{GL}{gl}$	×	$\frac{gl}{gl}$	
Gametes (*n*) with crossing over in meiosis		GL, gl, Gl, gL		
Offspring genotypes (2*n*)	$\frac{GL}{gl}$ $\frac{gl}{gl}$		$\frac{Gl}{gl}$ $\frac{gL}{gl}$	
	Parental genotypes		Recombinant genotypes	
Phenotypic ratio	41.5% 41.5%		8.5% 8.5%	

Fig. 15.7 Test-cross experiment

CHROMOSOME MAPPING

The number of cross-overs which takes place between two pairs of alleles is called the **cross-over value (COV)** and can be used to map the positions of genes on a chromosome. Maps are produced by converting COVs into arbitrary distances between genes. Suppose genes P and Q have a COV of 10% and genes P and R a COV of 6%, these genes would be 10 and 6 units apart respectively on the chromosome. Given this information there are two possible orders in which they could occur, these being P(10)Q(6)R or P(4)R(6)Q. As more information becomes available and further genes in the linkage group are ascertained it will become possible to decide between these two possibilities. For example, if gene S is found to be 4 units from P, 6 from Q and 12 from R, the definitive map would be P(4)S(6)Q(6)R. The bracketed numbers represent the distance between successive genes on the chromosomes.

SEX DETERMINATION AND LINKAGE

Most sexually reproducing animals show two morphologically distinct types, male and female, which are associated with the chromosomes found in the two types. In many insects, mammals and birds, the chromosome difference is that one pair, known as the **sex chromosomes** or **heterosomes**, are similar in one sex and dissimilar in the other. The other pairs of chromosomes are known as **autosomes**.

- In dipterans such as *Drosophila* and mammals, the male has dissimilar chromosomes, usually called X and Y, whilst the female has two similar X chromosomes.
- In birds and Lepidoptera the situation is reversed, the female being XY and the male XX.
- Sometimes in birds, these chromosomes are designated W and Z, the female being WZ and the male being WW.
- Another variation on the theme is that in some insects, e.g. grasshoppers, the males do not have a Y chromosome but only an X, thus the female is XX and the male XO.

If you draw a genetic diagram using XX female × XY male as the parental genotypes you will find that the expected sex ratio in the offspring is 1 female (XX): 1 male (XY).

- the sex which is XX is described as being **homogametic**: that is, producing similar gametes
- the sex which is XY is said to be **heterogametic**

SEX-LINKED INHERITANCE

The Y chromosome is much smaller than the X in most species and it carries very few, if any, genes. Therefore, in the heterogametic sex any recessive genes carried on the X chromosome will express themselves in the phenotype. This is because they are unpaired and so there is no dominant gene present. This special form of inheritance is known as **sex linkage**. An important feature of sex linkage is that a male cannot hand on the gene to his sons as they must receive the Y chromosome to be male. On the other hand, all his daughters must receive the recessive gene from him. Females who are heterozygous for sex-linked recessive traits are known as **carriers** and have a 50% chance of handing on the recessive to their sons. To obtain an affected (homozygous recessive) female, the father must be affected and the mother either affected or a carrier. The process of sex-linked inheritance is shown in Fig. 15.8. Examples of sex-linked traits are:

- white eye in *Drosophila*

and in humans:

- red/green colour blindness
- haemophilia
- Duchenne muscular dystrophy

> Do not make the mistake of thinking that sex-linked characters only affect the heterogametic sex

Parental phenotypes	Normal female	×	Colour-blind male		Carrier female	×	Normal male	
Parental genotypes (2n)	XX		X^cY		X^cX		XY	
Gametes (n)	X	X	X^c	Y	X^c	X	X	Y
F_1 genotypes (2n)	XX^c	XX^c	XY	X^cY	X^cX	XX	X^cY	XY
F_1 phenotype	Carrier females		Normal male	Colour-blind male	Carrier female	Normal female	Colour-blind male	Normal male

Parental phenotypes	Carrier female	×	Colour-blind male		Colour-blind female	×	Normal male	
Parental genotypes (2n)	X^cX		X^cY		X^cX^c		XY	
Gametes (n)	X^c	X	X^c	Y	X^c	X^c	X	Y
F_1 genotypes (2n)	X^cX^c	XX^c	X^cY	XY	X^cX	X^cX	X^cY	X^cY
F_2 phenotype	Colour-blind female	Carrier female	Colour-blind male	Normal male	Carrier females		Colour-blind males	

Fig. 15.8 Sex-linked inheritance red-green colour blindness in humans

MUTATION

A mutation is a change in the DNA of an organism; it may affect a single gene or a whole chromosome. Mutations occur randomly: that is, any gene can undergo mutation at any time. Mutations are caused by:

- **ionising radiation**
- **mutagenic chemicals**

Ionising radiation
The most important radiations are high-energy electromagnetic waves such as X-rays and gamma-rays and also high-energy particles such as α and β particles and neutrons. The majority of mutations are caused by these radiations.

Mutagenic chemicals
Mutagenic chemicals include mustard gas, benzene and its derivatives and formaldehyde.

CHROMOSOME MUTATIONS

These may affect either the number or the structure of chromosomes.

Changes in numbers

This is usually the result of errors occurring during meiosis. The loss or gain of a single chromosome is known as **aneuploidy**. An increase by a whole haploid set of chromosomes is known as **polyploidy**, which was described previously.

Aneuploidy results in the formation of gametes with either one too many or one too few chromosomes. One pair of chromosomes has failed to separate at anaphase and so the pair has moved to one pole of the cell, the phenomenon of **non-disjunction**. After fertilisation a zygote formed with an extra chromosome is said to be **trisomic**. Well-known trisomies in humans are Klinefelter's syndrome (XXY) and Down's syndrome (trisomy 21). A zygote with a chromosome missing is said to be **monosomic**. Most of these are inviable but Turner's syndrome (XO) in humans is viable, giving rise to an abnormal female.

Changes in structure

If mutagenic agents cause chromosomes to break, they may rejoin incorrectly.

- **Deletion**   A small piece of chromosome is lost.
- **Inversion**   A piece may be reversed before rejoining.
- **Translocation**   Breaks occur in two non-homologous chromosomes and parts of the two chromosomes are exchanged.

These chromosome mutations may cause phenotypic effects and often cause reduced fertility because they interfere with the process of meiosis.

GENE MUTATIONS

Any gene can mutate but rates vary from one gene to another within an organism. Gene mutations are changes in the base pairs within the genes. They can take the form of **duplication, insertion, deletion, inversion** or **substitution** of bases. Whatever the change, the result is the formation of a modified polypeptide. The bases are read off in triplets so any change causes a misreading.

For example, haemoglobin S is produced instead of normal haemoglobin by a single base change which causes valine to be substituted for glutamic acid at the sixth position in the β globin chain. DNA codes for glutamic acid are CTT or CTC. Two of the codes for valine are CAT and CAC. In either case the substitution of A for T as the second base would bring about the formation of haemoglobin S.

Some base changes will produce a nonsense or stop codes and so a polypeptide chain will be left incomplete.

POPULATION GENETICS

GENE POOL

A population is a group of organisms of the same species found within a clearly defined geographical area. Within the individuals of the population there are variations in phenotype and associated with these are variations in genotype. The total variety of genes present within a population is called the **gene pool**. If the environment is stable the gene pool will also be stable. However, if the environment is changing, some phenotypes will be at an advantage – that is, they will be selected – whilst others will be at a disadvantage and they will not be selected. In this situation the gene pool will be changing, some alleles becoming more frequent and others becoming less frequent. In some situations, alleles which are greatly disadvantaged may be totally lost from the gene pool.

ALLELE FREQUENCY

Two or more forms of a gene may exist and these are known as alleles. At a single locus on a chromosome, one or another allele of the gene found at that locus will be present.

CHAPTER 15 GENETICS AND EVOLUTION

When considering a whole population, each allele is found at a different frequency. As all the loci hold one, but not more, allele, the allele frequencies when added together must give 100%. Usually these percentages are stated as decimals and so for all the alleles of any gene, the total frequency must equal unity.

For example, in Europe the recessive allele for phenylketonuria (a biochemical defect in humans which leads to a mental defect if untreated) has a frequency of 1% or 0.01. Therefore, the dominant allele for the normal enzyme metabolism must have a frequency of 99% or 0.99.

THE HARDY-WEINBERG EQUATION

> This commonly occurs in questions. Be sure you understand the equation

The value of the Hardy-Weinberg equation is that it can be used to calculate allele and genotype frequencies in a population, assuming that the population is large and that there is no selection.

Using the equation

The basic premise is that if there are two alleles, the frequencies of the two remain constant in the population. Consider a pair of alleles Aa.

Let p = frequency of A, and q = frequency of a.
In the population:

$$p + q = 1$$

If the sexes are considered separately, the same equation applies.

In males:

$$p + q = 1$$

Similarly in females:

$$p + q = 1$$

Assuming random mating – that is, any male may mate with any female – the genotype frequencies in the next generation will be as follows:

$$AA = p^2; \; Aa = 2pq; \; aa = q^2$$

These frequencies arise because males $(p + q) \times$ females $(p + q)$ when expanded give the following :

$$(p + q)(p + q) = p^2 + 2pq + q^2$$

This equation is particularly useful for calculating the frequency of heterozygotes in a population – that is, the individuals in the population who show the dominant phenotype but carry the recessive allele.

GENETIC CHANGES IN POPULATIONS

Few, if any, natural populations show the conditions necessary for the Hardy-Weinberg equilibrium to exist, as sources of genetic variation are always present. Mutations of either genes or chromosomes take place and in most populations sexual selection occurs – that is, mating is non-random. Sometimes, variation in gene frequencies in populations occurs by chance. This is known as random **genetic drift** or the **Sewall Wright effect** and it may be an important evolutionary mechanism in small or isolated populations. The gene pool of a small isolated population is usually different from that of the population from which it originated. This is because the original members of the small population did not contain a representative sample of the alleles in the parent gene pool. This phenomenon is known as the **founder principle**. When two populations of a species are in contact with each other, mating takes place across the boundaries and so there is an exchange of genes, or gene flow, between the populations.

ORGANIC EVOLUTION

The concept of organic evolution has been around for over 2,000 years. In this time, evidence has accumulated and ideas have changed.

NATURAL SELECTION

The basis of present thought about organic evolution is the theory of evolution by natural selection put forward independently by Alfred Wallace and Charles Darwin. The theory is based on these observations:

- individuals within a population have the potential to produce large numbers of offspring
- the number of adults tends to stay the same
- variation is found in all populations

From these observations, two deductions were made:

- there is a struggle for existence, many individuals failing to survive
- the individuals best adapted to their environment survive and breed

As a result of the struggle for existence, natural selection takes place. The key to survival is adaptation to the environment. Variations, no matter how small, which give one individual an advantage over another confer a **selective advantage** and increase in number. Variations which are not favourable are eliminated from a population; they are subject to **selective disadvantage**.

NEO-DARWINISM

Since Darwin published his theory, much more evidence for evolution has been accumulated and the mechanisms of inheritance have been discovered. As a result, the basic theory has been extended and refined, using evidence from genetics and molecular biology along with a further hundred years' accumulation of evidence from palaeontology, ecology and behavioural studies. Particularly important is the light thrown on the mechanism of evolution by Mendelian genetics, the studies of population genetics and an understanding of the processes of specialisation which have arisen from it. This extension of Darwinian theory to include the genetic processes involved in natural selection is known as neo-Darwinism. Its basis is the production of evolutionary change by the accumulation of small genetic changes. Some biologists believe that such small changes cannot account for the appearance of new species but that these arise as a result of sudden large genetic changes in a stable population. However, there is a little present evidence for such changes having occurred.

TIME SCALE OF EVOLUTION

It is generally agreed that the Earth is 4.5–5.0×10^9 years old, but that life could not originate until the surface had cooled and solidified. The first known fossils are dated at 3.2×10^9 years old, which means that life had developed at some time prior to that.

- The first known organisms are very simple, with a structure similar to that of blue-green algae.
- Fossil burrows dated 7×10^8 years old have been found, similar to those formed in rocks by some present-day annelids, so by this time animal life was well established.
- By the end of the Cambrian period the whole range of invertebrate animals and many different algae are known.
- In the Devonian (3.5×10^8 years ago) all the groups of fish were established, along with the first reptiles and vascular plants.

- By the Jurassic (1.7×10^8 years ago) the mammals were established, there was a great variety of gymnosperms and the first angiosperms had appeared.
- During the Tertiary period mammals, birds and insects became the most important animals and angiosperms became the dominant land plants.
- Humans appeared later, maybe 1×10^7 years ago.

EVIDENCE FOR EVOLUTION

In Darwin's time the evidence consisted of fossils and data from comparative anatomy and selective breeding. Now a considerable amount of physiological and molecular biological evidence has been added to the information pointing to an evolutionary process.

Fossil evidence

A fossil is any preserved remains of prehistoric organisms. The commonest were formed where hard parts of organisms were immersed in water and became petrified. Sometimes organisms were embedded in mud, leaving a **mould** which subsequently became filled with sediment to form a **cast**. Occasionally whole organisms are preserved, in which case the soft anatomy can also be studied. Fossils on their own do not prove evolution but they show the appearance and disappearance of organisms over time.

- **Fossil dating** Fossils can be dated by dating the rocks in which they are found. The age of rocks is found by the technique of **radioisotope dating**. For fossils up to 50,000 years old, the decay of carbon^{14} → carbon^{12} gives an accurate figure. Older fossils are dated by the decay of uranium^{238} → lead^{206} or that of potassium^{40} → calcium^{40} and argon^{39}.

Geographical distribution

There are many examples of modern, closely related animals being found in widely separated areas: for example, llamas in South America, dromedaries in Africa and Bactrian camels in Asia. In each case fossil evidence shows a much wider distribution and a common ancestry. From their origin, organisms migrated over land bridges or the original population was split by continental drift. The existence of the monotreme and marsupial population of Australia is explained by this area breaking away from the mainland areas in the late Jurassic, before any placental mammals were present.

Comparative anatomy

When the anatomy of groups of apparently unrelated plants or animals is compared, similarities appear which suggest the possibility of evolution from a common ancestor. Examples are:

- the similarity of basic structure in all flowers
- the common bone pattern, **pentadactyl limb**, of all tetrapod vertebrates

Organs with a similar structure and embryological development are said to be **homologous**. The modifications which have taken place to change the basic structure into specialisations for specific functions are known as **adaptive radiation** or **divergent evolution**. The existence of non-functional **vestigial organs** which are homologous with functional organs in related species also gives evidence of evolution from a common ancestor. Sometimes similar structure or physiological processes are said to be **analogous** and represent **convergent evolution**. Examples include:

- the wings of bats and insects
- the eyes of cephalopods and vertebrates

Comparative physiology and biochemistry

The composition of body fluids is similar in all animals and it resembles the composition of the seawater in which life is thought to have originated. Many enzymes are common to all animals, from protozoa through to vertebrates, and pigments such as haemoglobin and cytochrome are found in many different animals with only minor

> This is the important recent evidence

differences between species. Cytochrome c, for example, has a polypeptide chain varying from 104 to 112 amino acids depending on species. Cytochrome c in humans and chimpanzees is identical. The minor changes in cytochrome c can be used to chart evolutionary relationships.

Immunological research has shown relationships not only between primates but also between other groups: for example, between the king crab (*Limulus*) and arachnids. As a result of this research, *Limulus* is now classified as an arachnid.

SELECTION

Darwin's observations of variation within a population and the tendency for the adult population to be stable in size led to the development of the idea of **natural selection**. Environmental factors influence survival. In any particular environment, some genetic changes will lead to advantageous genotypes and these will survive. The accumulation of advantageous changes is thought to lead to the development of new varieties, subspecies and eventually to new species. This is the process of natural selection. Natural selection is a slow process and usually cannot be observed, an exception being the appearance of a melanic form in some moth species – for example, *Biston betularia* – which appeared after the industrial revolution.

SELECTIVE BREEDING

Selective breeding of animals and plants makes use of variations which occur within a population. In this case, people choose organisms showing desirable characters and breed only from these. This process of **artificial selection** mimics natural selection and provides evidence that selection can lead to the development of characteristics and the production of very distinct forms of organisms, as seen in many domestic animal and plant species.

ORIGIN OF SPECIES

New species arise when some barrier to reproduction occurs so that the gene pool is divided and different combinations of mutants develop in each pool division. If the separation is long term, eventually the two groups are so different that two new species incapable of interbreeding are formed. For new species to develop from a population, some form of isolating mechanism is required. Such mechanisms may prevent the formation of either hybrids or their reproduction if formed.

GEOGRAPHICAL ISOLATION

Geographical isolation occurs when a physical barrier divides a population into two or more parts. The barrier could be a mountain range, the breaking of a land bridge, continental drift or, on a smaller scale, the widening of a river.

ECOLOGICAL ISOLATION

Ecological isolation occurs when two species inhabit the same region but have different habitat preferences.

SEASONAL ISOLATION

Seasonal isolation occurs when two species mate or flower at different times of the year.

BEHAVIOURAL ISOLATION

Behavioural isolation occurs when courtship behaviour differs so that mating between two species does not take place.

MECHANICAL ISOLATION

Mechanical isolation occurs where differences in the reproductive organs prevent successful copulation.

All these mechanisms prevent the formation of hybrids between species.

Mechanisms which allow the formation of hybrids but which prevent them from producing offspring include:

- **hybrid inviability**, where the hybrids formed fail to reach maturity
- **hybrid sterility**, where the hybrids fail to produce viable gametes
- **hybrid breakdown**, where the F_1 hybrids are fertile but the F_2 generation and offspring of back crosses to the parents are sterile.

All these mechanisms prevent interbreeding between species, but the species would possibly be formed in the first place by a geographical barrier separating the original gene pool. Different mutants would accumulate in the subdivisions as each group adapted to its separate environment and if the separation were long enough the groups would be genetically isolated when brought together again: that is, they would be different species. Good examples of this are:

- Darwin's finches in the Galapagos
- honey creepers in the Hawaiian islands

CONSERVATION OF GENE POOLS

Human activities since the beginning of settled agriculture and the development of urban living have had various deleterious effects on the environment and on the existence of 'naturally' occurring animals and plants. These activities have affected gene pools in two ways.

EXTINCTION OF SPECIES

The destruction of natural habitats and the excessive hunting of animals have together led to the extinction of many animal and plant species, with the consequent complete loss of genetic material. Well-known examples of this are:

- the extinction of the dodo from Mauritius
- the extinction of the passenger pigeon from North America
- the extinction of the quagga (a zebra species) from South Africa

In recent years there has been much concern about this loss of gene pools and various legislation has endeavoured to prevent the extinction of endangered species. Along with this, zoos worldwide have become involved in breeding endangered species and in many cases stock has been released in the original wild habitats. An example of this is the release of Arabian oryx in Oman in 1982 and in Jordan in 1983.

BREEDING OF DOMESTIC ANIMALS AND PLANTS

The other main effect has been in the breeding of domestic animals and plants. By its nature, selective breeding leads to changes in the proportions of genes in the gene pool and to the eventual loss of some genes. In the course of time, therefore, there is a reduction in the variability of these species. This means that future possibilities for change may be limited and the long-term survival of species consequently may be jeopardised. As a result, steps have been taken to preserve as wide a gene pool as possible for domestic species. Stocks of seeds of 'traditional' varieties of plants are stored, rare breed societies maintain old, less commercial varieties of animals and in some specialised zoos attempts are being made to breed to earlier varieties of animals.

PRACTICAL WORK

Experimental work on monohybrid and dihybrid crosses and linkage can be carried out using *Drosophila*, mice, tomatoes and other organisms.

Mitosis and meiosis should be studied practically. Mitotic stages can be seen most easily in onion root tips. The tip, about 2 mm long, should be removed and mounted on a slide with a chromosome stain such as aceto-orcein. A cover slip should be placed over the tip, which is then squashed. Meiosis can be seen in sections of anthers or squashes of locust testis.

Perhaps the most important aspect of coursework in genetics is to practise answering genetics problems. Solving these gives a thorough understanding of the elements of genetics. When solving problems it is essential to reason logically from what you are told to the solution. Write down each step in the reasoning. In examinations, most of the marks are for the steps in the solution, not the final answer.

Visits to museums and the use of film and videotape will help your knowledge of evolution and the evidence for it. Similarly, visits to specialist zoos or rare breed society farms will be invaluable for an understanding of the work involved in conserving gene pools.

Further reading
Berry R J, 1982, *Neodarwinism*. Arnold
Bowman J C, 1981, 'The conservation of farm animals', *Biologist* 28 (4) 116–92
Gardner E J, 1981, *Principles of genetics* (6th edn). Wiley
Lawrence W J C, 1968, 'Survival reservoirs for endangered species: the conservation role of a modern zoo', *Biologist* 31 (2) 79–84

EXAMINATION QUESTIONS AND ANSWERS

QUESTIONS

Question 15.1
a) If you were conducting breeding experiments with a population in which every individual is heterozygous for a particular gene, what would you conclude about the alleles of this gene if the offspring showed:
 i) only two phenotypes
 ii) only three phenotypes
 iii) a large number of phenotypes?
b) It is said that a person's ability to smell Freesia flowers is genetically controlled. How could you investigate this hypothesis?
(6)
(Total 6 marks)
(AEB)

Question 15.2
Explain what is meant by **four** of the following:
a) homologous chromosomes
b) linkage group
c) monohybrid and dihybrid ratios
d) genes
e) mutagens
(4 × 5)
(Total 20 marks)
(ULEAC)

Question 15.3
Explain why each of the following statements is wrong:
a) Cross-fertilisation enlarges the gene pool. *(3)*
b) Offspring resulting from self-fertilisation are genetically identical. *(3)*
c) Different phenotypes are the result of different genotypes. *(3)*
(Total 9 marks)
(AEB)

CHAPTER 15 GENETICS AND EVOLUTION

Question 15.4
In humans, the condition of cystic fibrosis is caused by an allele a which is recessive to the normal allele, A.

a) Complete the spaces below and show the probability of two people, both heterozygous for this condition, having a child with cystic fibrosis.

	Mother	Father
Parental phenotypes	_____	_____
Parental genotypes	_____	_____
Gamete genotypes	_____	_____
Offspring genotypes	_____	_____
Probability of offspring having cystic fibrosis		_____

(2)

b) In Britain, approximately 1 in 25 people are heterozygous for this condition. What is the probability that any two parents are both heterozygotes? *(1)*

c) The frequency of the cystic fibrosis allele in Britain is 0.02. Calculate the expected frequency of people who are:
  i) homozygous aa;
  ii) homozygous AA.
  In each case, show your working. *(3)*

(Total 6 marks)
(AEB)

Question 15.5
a) What is a fossil? *(1)*
b) Describe ways in which a fossil may be formed. *(2)*
c) Explain the following statements:
  i) Whole mammoths found in Siberia and the leaves of the arctic plant *Dryas* found in post-glacial peats in Southern England do not qualify as fossils. *(3)*
  ii) Worms and flowers are only rarely found as fossils. *(2)*
  iii) Fossil remains of early man are comparatively rare. *(2)*
  iv) Fossils of aquatic animals are more abundant in the earth's crust than fossils of terrestrial animals. *(2)*

(Total 12 marks)
(ULEAC)

OUTLINE ANSWERS

Answer 15.1
a) i) Two alleles, one dominant, one recessive
   ii) Two alleles, co-dominant/incompletely dominant
   iii) Several multiple alleles
b) Use several families; test all members; construct family trees showing those who can and cannot smell *Freesia*; any genetic relationship will become clear.

Answer 15.2
a) Pairs of chromosomes; one from maternal gamete, one from paternal; chromosomes identical in length; have the same number of genes; centromeres in the same position.
b) All the genes on one chromosome; number of linkage groups same as number of chromosomes; alleles of linked genes show little assortment; inherited together unless separated by crossing over; majority of offspring show parental phenotypes but a few show new combinations.
c) Monohybrid inheritance involves one pair of alleles; ratio of 3 dominant: 1 recessive in F_2 generation; dihybrid inheritance involves two pairs of alleles; ratio of 9:3:3:1 in F_2; genes on different chromosomes, so random assortment.
d) Determine an organism's characteristics; capable of replication; located on chromosome; coding for a polypeptide chain; consists of a section of DNA.
e) Substances which cause mutations; mutation is a change in the DNA or in the structure of a chromosome; chemicals such as mustard gas; high-energy electromagnetic radiation, e.g. X-rays, cause changes in a characteristic if they affect gamete nuclei.

Answer 15.3

a) Cross-fertilisation reassorts genes; new combinations and possible new phenotypes produced; but no increase in the number or proportions of the genes within the population (gene pool).
b) Self-fertilisation results in new gene combinations as does cross-fertilisation; offspring would be genetically identical if parent was homozygous at all loci; this is very unlikely to occur.
c) The same phenotype may be produced by different genotypes; homozygous dominant and heterozygous genotypes give the same phenotype; the same genotype can produce different phenotypes due to environmental influences; e.g. Himalayan rabbits.

Answer 15.4

a)
	Mother	Father
Parental phenotypes	normal	normal
Parental genotypes	Aa	Aa
Gamete genotypes	A, a	A, a
Offspring genotypes	AA, Aa, Aa, aa	
Probability of offspring having cystic fibrosis 25%		

b) 1 in 625
c) i) homozygous aa = $0.02 \times 0.2 = 0.0004$
   ii) frequency of normal allele = $1 - 0.02 = 0.98$
       homozygous AA = $0.98 \times 0.98 = 0.96$

Answer 15.5

a) Indication of or remains of a living organism present in the distant past which are preserved in rock.
b) Petrification, parts replaced by minerals; cast of organism encased by rock; impression of parts in rock; whole organism trapped in resin, which becomes amber (any two).
c) i) No change in form, original tissues still present; not in rock; mammoths deep-frozen; *Dryas* in peat.
   ii) No hard tissues, e.g. bone or lignin; decay too quickly for fossil to form.
   iii) Populations very small; did not live in places where fossilisation easily occurs.
   iv) Sedimentation rapid in estuaries and beaches; much larger populations of aquatic animals; greater area covered by water.

QUESTIONS, STUDENT ANSWERS AND EXAMINER COMMENTS

Question 15.6

In a genetics experiment, tomato plants which were heterozygous for the recessive allele non-pigmented (green) stem and smooth ('entire') leaflets were allowed to interbreed.

Three hundred and twenty seedlings were produced as a result of the cross and these were sorted into four different phenotype groups and counted. The results were:

Pigmented (purple) stem, serrated ('cut') leaflets	190
Pigmented (purple) stem, smooth ('entire') leaflets	53
Non-pigmented (green) stem, serrated ('cut') leaflets	63
Non-pigmented (green) stem, smooth ('entire') leaflets	14
Total	320

a) Fill in the necessary figures in the table below that would be required in order to calculate the χ^2 value.
   Assume that the inheritance is of simple Mendelian type giving a 9:3:3:1 ratio of phenotypes.
   Purple (P) is dominant to green (p) and serrated (S) is dominant to smooth (s).

(5)

	P/S	P/s	p/S	p/s
Observed numbers	190	53	63	14
Expected numbers	180	60	60	20
O–E	10	–7	3	–6
$(O-E)^2$	100	49	9	36
$\dfrac{(O-E)^2}{E}$	0.56	0.81	0.15	1.8

$\chi^2 = 3.32$

> **Correct**

b) In this example the number of degrees of freedom is 3. Below is an extract from χ^2 distribution tables giving the probability levels for 3 degrees of freedom.

Probability	0.99	0.95	0.05	0.001
χ^2	0.115	0.352	7.82	11.35

i) Explain briefly how the experimenter could interpret this χ^2 value of 3.32 and what conclusions could be drawn.

The value gives a probability between 0.95 and 0.95, being slightly nearer 0.95. The hypothesis has not conclusively been either supported nor proved incorrect. (3)

> **Not a good answer, you cannot speculate on the level of probability. Acceptable as probability is greater than 0.05; hypothesis accepted; the data agrees with a 9:3:3:1 ratio**

ii) If a χ^2 value of greater than 7.82 (probability is less than 0.05) had been found, what conclusion would you have drawn?

That the null hypothesis of a 9:3:3:1 ratio was not supported by these results. (1)

> **Correct**

(Total 9 marks)
(ODLE)

Question 15.6

A gardener regularly saves seed from his crop of beans for sowing the following year. Occasionally plants were obtained which flowered but produced no seeds, though they were cross-pollinated in the usual way. It was suspected that this was caused by a recessive allele, a, which prevents seed production.

a) make a simple genetic diagram to show how sterile plants could appear in the population. Use A, a, to represent the alleles.

```
P          Aa         ×          Aa
gametes    A,a                   A,a
F₁         AA         Aa         Aa         aa
                      |
                 fertile plants —           sterile plants —
                 produce seed               no seed produced
```
(4)

> **Correct. As much detail as required**

b) Give **two** reasons why the gardener only occasionally finds sterile plants.

i) *a remains in population only in the heterozygote*

ii) *few aa plants are produced* (2)

> **Correct**

> **Repeats the question. A better answer would be: 1 in 4 chance of a sterile plant when heterozygotes cross**

c) Using genetic diagrams, suggest how the gardener could eliminate the allele a from his bean population during controlled crossings. (The beans cannot be self-pollinated.) Indicate clearly the unsuitable crosses as well as the number of generations required to ensure allele a has been eliminated.

Cross 1   P   AA × AA

> **Correct**

F^1 AA
F^2 AA → no heterozygotes produced

Cross 2   P   AA × Aa
F^1 AA or Aa
F^2 AA1 Aa and aa produced some sterile.

Cross 3   P   Aa × Aa
F$_1$ AA, Aa, Aa, aa some sterile

> This statement is correct but an F$_3$ generation should have been put in to prove that all the F$_2$ are homozygous AA, i.e. four generations are needed

> Again should have stated seed not used

(8)
(Total 14 marks)
(ULEAC)

REVIEW SHEET

1 Define each of the following.
   a) Discontinuous variation: _____
   _____
   _____

   b) Continuous variation: _____
   _____
   _____

   c) Mitosis: _____
   _____
   _____

   d) Meiosis: _____
   _____
   _____

2 Complete the labels for each diagram then describe the key characteristics of that *phase* within **mitosis**.

   a) Interphase: _____
   _____
   _____
   _____
   _____
   _____

   b) Prophase: _____
   _____
   _____
   _____
   _____
   _____

   c) Metaphase: _____
   _____
   _____
   _____
   _____
   _____

   d) Anaphase: _____
   _____
   _____
   _____
   _____
   _____

e) Telophase: _____
_____
_____
_____
_____
_____
_____

3 Draw an appropriate diagram and then use it to explain each of the stages of **Meiosis I**.

a) Prophase I: _____
_____
_____

b) Metaphase I: _____
_____
_____

c) Anaphase I: _____
_____
_____

d) Telophase I: _____
_____
_____

4 What is involved in the **second meiotic division**:
_____
_____
_____

CHAPTER 15 REVIEW SHEET

5 a) Define **Mendel's first law**. _____
_____
_____
_____

b) Complete the **monohybrid cross** shown below.

T represents the allele for tall plants (dominant)
t represents the allele for dwarf plants (recessive)
2n is the diploid state with pairs of chromosomes
n is the haploid state with single chromosomes

Parental phenotypes	*Pure breeding tall plants*		*Pure breeding dwarf plants*	
Parental genotypes (2n)	☐	×	☐	
Gametes	☐	☐	☐	☐
F₁ genotype (2n) by random fertilisation	☐	☐	☐	☐
	F₁ genotypes all ☐			
	F₁ phenotypes all ☐			
Self-pollination of the F₁ plants	*Tall plant*		*Tall plant*	
F₁ genotypes (2n)	☐	×	☐	
Gametes (n)	☐	☐	☐	☐
F₂ genotypes (2n) by random fertilisation	☐	☐	☐	☐
	F₂ phenotypes ☐ tall : ☐ dwarf			

6 a) Define **Mendel's second law**. _____
_____
_____
_____

b) Complete the **punnett square**.

R represents round seed (dominant)
Y represents yellow seed (dominant)
r represents wrinkled seed (recessive)
y represents green seed (recessive)

Parental phenotypes	Pure breeding round yellow			Pure breeding wrinkled green
Parental genotypes (2n)	RRYY			rryy
Gametes (n)	all RY			all ry
F₁ genotype (2n) random fertilisation		all RrYy		
Self fertilisation		RrYy × RrYy		
Gametes identical from ovules and pollen grains	RY	Ry	rY	Yy

F₂ genotypes (2n) and phenotypes shown in a Punnett square

round yellow
round green
wrinkled yellow
wrinkled green

♀ \ ♂				

7 What do you understand by **incomplete dominance**?

_____
_____
_____
_____

8 a) What is meant by **gene interaction**?

_____
_____

b) Complete the punnett square below to illustrate gene interaction leading to comb shape in domestic hens.

R represents rose comb (dominant)
r represents single comb (recessive)
P represents pea comb (dominant)
p represents single comb (recessive)

Parental phenotypes	Pure breeding rose comb		Pure breeding pea comb
Parental genotypes (2n)	RRpp	×	rrPP
Gametes (n)	all Rp		all rp
F_1 genotype (2n) random fertilisation		all RrPp	
F_1 phenotype (2n)		all walnut comb	
cross fertilisation F_1		RrPp × RrPp	

Gametes identical for eggs and sperm

F_2 genotypes (2n) and phenotypes shown in a Punnett square

walnut
rose
pea
single

♀ \ ♂				

9 Briefly outline the **chromosomal theory of inheritance**.

_____
_____
_____
_____
_____

10 What do you understand by **chromosome mapping**?

_____
_____
_____
_____
_____

GETTING STARTED

Ecology forms an important part of all syllabuses. A major aspect is the study of one or more specific ecosystems. Some syllabuses specify which should be studied but others leave an open choice. Questions set are sometimes about an ecosystem which you have studied in the field. To answer these questions, you must have carried out a thorough investigation, using quantitative methods such as **transects** or **quadrats** and have analysed and interpreted your results. Often you will be asked to explain techniques you have used and to quote summary data and relationships which you have found. Questions based on the interpretation of ecological data are also common.

It is important to be familiar with ecological terminology to avoid confusion between terms; for example, **population** and **community** or **microhabitat** and **niche**. An understanding of ecological dynamics is also important to help you to interpret your own practical findings.

You should use this chapter along with Chapter 3, Diversity of Organisms, and also the physiological chapters, as questions often involve cross-reference between physiological processes and the conditions in which organisms live.

There is usually a fieldwork component in the course assessment and for this you may have to carry out a long-term field investigation. In this context investigations spanning more than one season are valuable, giving good opportunities for interpretation and comment.

CHAPTER 16
ECOLOGY

ECOSYSTEMS

ENERGY FLOW AND NUTRIENT CYCLES

FOOD CHAINS AND FOOD WEBS

ECOLOGICAL PYRAMIDS

POPULATION GROWTH AND CONTROL

SUCCESSION AND CLIMAX

AUTECOLOGY AND SYNECOLOGY

POLLUTION OF ECOSYSTEMS

CONSERVATION

PRACTICAL WORK

CHAPTER 16 ECOLOGY

ESSENTIAL PRINCIPLES

ECOSYSTEMS

An **ecosystem** is a community of organisms which, together with their physical environment, form a self-perpetuating ecological unit. Usually the term ecosystem is used for a clearly defined area with distinctive flora and fauna. One system may overlap other systems at its edges and so, although one system is an entity in itself, its boundaries may be difficult to define.

COMPONENTS OF AN ECOSYSTEM

- A **population** is a group of organisms of a single species occupying a particular area. In this area there will also be other populations, forming a **community**.
- **Habitat** is the particular area occupied by a population. It has **biotic** and **abiotic** features which separate it from other habitats.
  - The biotic features are the sum total of the organisms within the habitat and their interactions.
  - Abiotic features include different types.
    - **Edaphic** features relate to the soil and include all its physical and chemical characteristics.
    - **Climatic** features include light, temperature, moisture, salinity and, particularly, the stability or variability of these.
- **Microhabitats** are areas of varying characteristics within a habitat.
- **Ecological niche** is the place of each species in an ecosystem. This is not only the physical space which it occupies, but the role which it carries out within the community and its interrelationships with other species as well. In the long term, two species cannot occupy the same niche in a specific habitat. However, in similar habitats in different parts of the world, species often taxonomically very different have evolved to occupy comparable niches. The relationships between the components of an ecosystem are shown in Fig. 16.1.

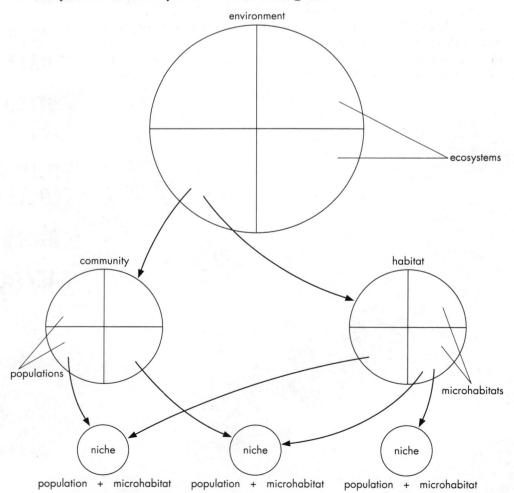

Fig. 16.1 Relationship between components of an ecosystem

ENERGY FLOW AND NUTRIENT CYCLES

ENERGY SOURCE

The ultimate source of energy for ecosystems is the sun, where energy is released in the form of electromagnetic waves. Approximately half the total radiation is in the form of visible light. In a year, the total amount of energy entering the earth's atmosphere is about 64.26×10^8 kJ m^{-2}, but only a small part of this energy enters ecosystems. Also the quantity absorbed by plants varies considerably at different latitudes. In Britain the average amount per year is 10.5×10^8 kJ m^{-2}. Most of the energy absorbed by green plants is in the red and blue regions of the visible spectrum. Of the energy entering a plant, only 1% to 5% is used; the rest is lost, partly by reflection and partly by the evaporation of water.

ENERGY FLOW

> Remember that energy is not a cycle

Energy enters ecosystems via photosynthesis and is fixed in the form of organic compounds. These pass to heterotrophs when plants are eaten by animals and to decomposers when animals and plants die. At each stage organic material is oxidised to release energy for metabolic processes and some of the energy is dissipated as heat and so is lost to the system. Constant new supplies of energy are needed to maintain the system. This energy flow is summarised in Fig. 16.2. Energy constantly enters and leaves an ecosystem but nutrients are constantly recycled.

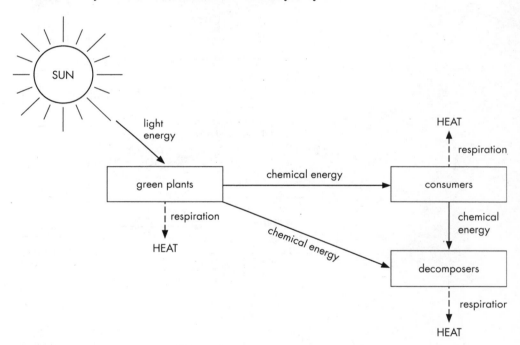

Fig. 16.2 Energy flow in an ecosystem

BIOGEOCHEMICAL CYCLES

Biogeochemical cycles are cycles by which chemical elements pass from the environment through a sequence of living organisms and back into the environment. The cycles most important to life are those of carbon, hydrogen, nitrogen, phosphorus, oxygen and sulphur. The carbon and nitrogen cycles are specifically featured in A-level syllabuses but the cycling of oxygen is implied via the work on photosynthesis and respiration.

Nitrogen cycle

The nitrogen cycle is the flow of inorganic and organic nitrogen within an ecosystem where there is an interchange between nitrogenous compounds and atmospheric nitrogen. The main processes involved are as follows.

- **Putrefaction** Decay processes convert organic nitrogen into ammonia. Various bacteria, actinomycetes and fungi carry out this process of putrefaction or **ammonification.**

- **Nitrification** The ammonia formed in putrefaction is converted by nitrification via nitrites to nitrates, the main absorbable form of nitrogen. Various bacteria are involved: for example, ammonia is converted to nitrite by *Nitrosomonas* and nitrite to nitrate by *Nitrobacter*, *Nitrocystis* and *Nitrococcus*.
- **Nitrogen fixation** Atmospheric nitrogen can be converted to nitrogen compounds by nitrogen fixation. Various free-living bacteria such as *Azotobacter* and some species of *Clostridium* are nitrogen fixers, as are some species of Cyanobacteria such as *Anabaena* and *Trichodesmium*. There are also symbiotic nitrogen fixers such as the bacterium *Rhizobium*, found in leguminous root nodules and the various mycorrhizal fungi.
- **Denitrification** Nitrogen is lost from ecosystems by denitrification. This is a particular problem in waterlogged soils with anaerobic conditions. Most denitrifiers either reduce nitrates to nitrites or nitrites to ammonia, but a few, such as the bacterium *Pseudomonas*, can reduce nitrates to molecular nitrogen. The nitrogen cycle is summarised in Fig. 16.3.

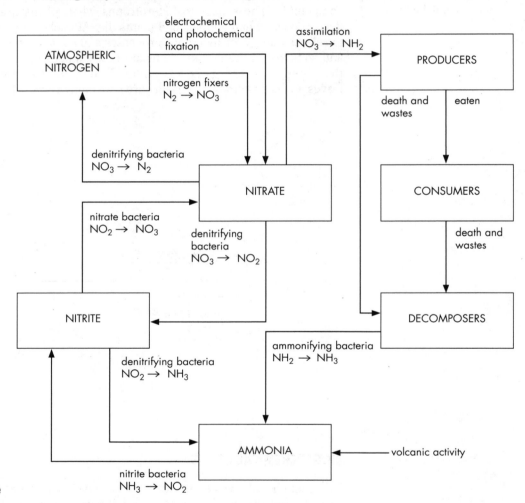

Fig. 16.3 Nitrogen cycle

Carbon cycle

The carbon cycle (Fig. 16.4) is the route taken by carbon from the atmosphere (as CO_2) after fixation in **photosynthesis** in green plants. The production of carbohydrates, proteins and fats contributes to plant growth and subsequently to animal growth through complex food webs. The dead remains of both plants and animals are then acted upon by saprophytes in the soil which ultimately release gaseous CO_2 back to the atmosphere. During life, CO_2 is also returned to the atmosphere through respiration, and indirectly through nitrogenous excretion.

Sulphur cycle

Sulphur is abundant in the earth's crust and is mainly available to plants as sulphate. Decay processes convert organic sulphur (in most proteins) to sulphides, which are then oxidised back to the sulphates used by plants.

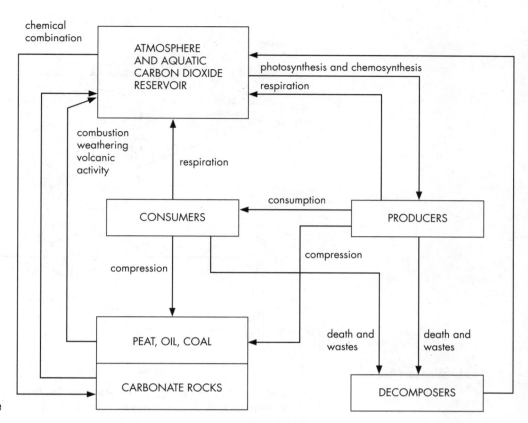

Fig. 16.4 Carbon cycle

Phosphorus cycle
Phosphorus is also cycled but much of it becomes locked up in sediments and so it is in relatively short supply, often being a limiting factor in ecosystems.

FOOD CHAINS AND FOOD WEBS

The organisms in an ecosystem have differing sources of food, ranging from green plants which manufacture organic matter (**producers**) through animals feeding on these plants (**herbivores** or **primary consumers**) to animals feeding on these animals (**carnivores** or **secondary** and **tertiary consumers**). Each of these groups forms a **trophic level**, with energy passing from each level to a higher one as material is eaten. At each level, energy is lost through respiration, waste products and unused material, so the amount of energy is reduced. The sequence from plant to herbivore to carnivore is a **food chain** and is the route by which energy passes between trophic levels. At the base of the chain are the **producers**. These are autotrophic and able to synthesise complex organic molecules using energy from the sun in most cases. Producers form the first trophic level, and they are fed upon by **primary consumers** on the second trophic level. These in turn are fed upon by **secondary consumers**, which comprise the third trophic level in the chain, and so on. For example,

grass → sheep → human being
PRODUCER   FIRST CONSUMER   SECOND CONSUMER

The number of links in the chain is normally limited to four or five. This is because although there is a flow of energy through the chain, a high percentage of the energy input is lost at each stage, through respiration, excretion and other body activities.

A **detritivore food chain** is slightly different from the one already described. It occurs where there is dead organic material which is fed on by decomposers. These in turn are food for secondary consumers.

Single food chains in fact rarely exist, and a **food web** is a more realistic representation of feeding relationships. This is because most primary consumers feed on more than one kind of autotroph, and most secondary consumers rely on more than one type of prey. The more varied the organisms in an environment, the more complex a food web is formed (the one illustrated in Fig. 16.5 is very simplified). Studies of feeding relationships have become extremely important in modern ecological research.

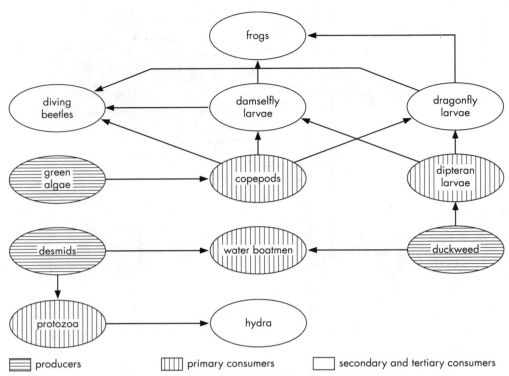

Fig. 16.5 Simplified pond food web

▭ producers   ▥ primary consumers   ▭ secondary and tertiary consumers

The original source of energy is the sun and the final loss of energy from a system is via **detritivores** and **decomposers**. The loss of energy between trophic levels is usually about 90%, so food chains rarely have more than four or five trophic levels.

ECOLOGICAL PYRAMIDS

The number of organisms, their biomass or the amount of energy contained in each trophic level can be represented in diagrams with a bar for each level. These are known as pyramids.

PYRAMIDS OF NUMBERS

A pyramid of numbers shows that the number of organisms at each trophic level usually decreases (i.e. most producers, less primary consumers, even less secondary consumers). This is because the loss of energy through the chain means that the higher levels can support fewer numbers and also the animals at the top of the chain tend to be larger anyway. There are exceptions, however, as when a single producer, such as a tree, supports large numbers of consumers, or a single host organism supports many parasites. Pyramids of numbers are therefore not very accurate representations of feeding relationships.

Fig. 16.6 shows pyramids of numbers:

A shows the typical pyramid, where the producers are small but found in large numbers.

B represents a situation often found in woodland ecosystems where the producer level consists of a few very large plants.

C is a specialised situation where the second and third levels consist of parasites and hyperparasites.

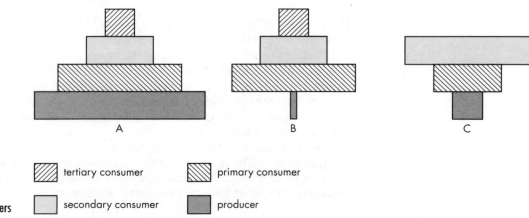

Fig. 16.6 Pyramids of numbers

PYRAMIDS OF BIOMASS

A pyramid of biomass reflects the decrease in biomass at each trophic level in a food chain. The biomass is the total weight of living matter, so it reflects both the numbers of organisms at each trophic level and also their size. It is therefore a clearer representation than a pyramid of numbers, but gives no indication of the relative energy contents of plant and animal tissues.

Fig. 16.7 shows pyramids of biomass. These are based on the mass of organisms found at each trophic level. The majority of pyramids are like **A**, where the total mass of each level is smaller than that of the level below. If the producers are short-lived, at any one time the mass of producers is smaller than that of the primary consumers. However, over a period of time the biomass of the producers would be greater. This situation is shown in pyramid **B**.

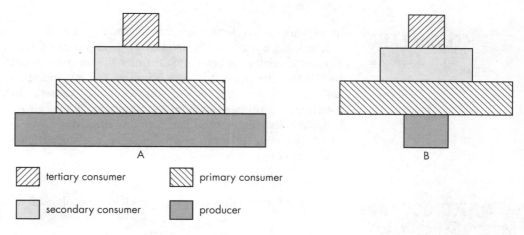

Fig. 16.7 Pyramids of biomass

PYRAMIDS OF ENERGY

> Be able to discuss the pros and cons of the different types of pyramid

A pyramid of energy represents the total energy content of each successive trophic level in a food chain. As material passes up through the food chain, energy is lost in respiration as heat so the sizes of bars decrease sharply. The efficiency of transfer of energy from plants to herbivores, for example, is only about 20 per cent. They are usually constructed for the energy utilised by the different feeding types in a unit area over a set period of time. An energy pyramid overcomes all the difficulties of comparisons of ecosystems which arise from the different sizes of organisms and the differences in energy equivalent to unit masses of tissue. A hypothetical pyramid of energy is shown in Fig. 16.8.

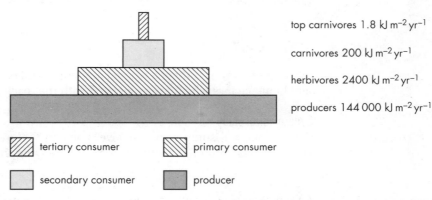

top carnivores 1.8 kJ m^{-2} yr^{-1}

carnivores 200 kJ m^{-2} yr^{-1}

herbivores 2400 kJ m^{-2} yr^{-1}

producers 144 000 kJ m^{-2} yr^{-1}

Fig 16.8 Pyramid of energy

POPULATION GROWTH AND CONTROL

Whether or not a population increases in size depends on:

- the relationships between birth rate and death rate
- the effect of migration into or out of the population

Many different factors affect these parameters: for example, available food, predation, overcrowding and disease. These factors mainly affect mortality but may affect birth rate or migration. Overcrowding, in particular, tends to reduce the birth rate and to induce outward migration.

DENSITY DEPENDENCE

- Some factors are **density dependent**: that is, their effect increases as the density of the population increases. Examples are the accumulation of toxic wastes, disease, parasitism and sometimes food supply.
- Other factors are **density independent**: that is, their effect does not depend upon the population density. Examples are migration, seasonal changes and changes in reproductive rate.

SUCCESSION AND CLIMAX

In any area, over time, new organisms replace existing ones until a stable state is reached. Bare rock is colonised by algae and lichens, forming a **pioneer community**. The accumulation of dead and decomposing organic material and the weathering of the rock lead to the formation of a primitive soil in which higher plants can grow. Mosses and ferns appear and as the soil develops, grasses, shrubs and trees appear. This constitutes a **succession** and eventually leads to a **climax community**. On land this is usually deciduous woodland with one **dominant** or several **co-dominant** species of trees. Human interference produces an unnatural climax, a **plagioclimax**, such as chalk grassland or lowland heath.

AUTECOLOGY AND SYNECOLOGY

> You must be able to define these terms

These are the two different ways of tackling ecological investigations.

AUTECOLOGY

Autecology is the study of a single species within an ecosystem. It is an investigation of all the factors which affect the species and of all the relationships it makes with other species in the community.

Autecological studies

An autecological study of an animal would include investigation of its classification, habitat, structure, movement and relationships with other species.

A similar study of a plant would include classification, habitat, structure, physiology, reproduction, life cycle, distribution within the habitat and relationship with animals in the ecosystem.

SYNECOLOGY

Synecology is the study of a whole community and consists of investigating the abiotic and biotic factors which are associated with that particular community.

Synecological study

A synecological study would involve mapping the area, identifying the species living there, investigating the abiotic factors and studying the biotic relationships.

POLLUTION OF ECOSYSTEMS

Of the various pollutants of land, water and air, the main ones specifically mentioned in syllabuses are sulphur dioxide as an air pollutant and nitrates as pollutants of freshwater.

SULPHUR DIOXIDE

Sulphur dioxide is formed when fossil fuels are burned. The high levels produced in industrial societies have damaging effects on plant life and so indirectly on animals.

Sulphur dioxide can be absorbed through stomata into leaves, damaging sensitive plants such as wheat and barley, but the main effects are from the acidification of soil and freshwater. The gas reacts with water to form sulphuric acid and falls as **acid rain**, lowering the pH of the soil and bringing into solution other toxic substances such as aluminium. The main effects are the destruction of vast areas of coniferous trees and the death of organisms in freshwater lakes in affected areas.

NITRATES

> It is important to remember that nitrate pollution accelerates not causes eutrophication

Eutrophication is the natural ageing of a freshwater lake. Pollution of freshwater by nitrates, mainly as run-off from arable land, accelerates the eutrophication process. Because of the influx of nutrients, blooms of algae appear. When they die they increase the amount of dead organic matter, which encourages the activity of decay bacteria, leading to an increase in the **biochemical oxygen demand** (**BOD**) and consequent deoxygenation of the water. This in turn kills the animal population. Large inputs of sewage or other nitrogen-rich material have a similar effect.

OTHER POLLUTANTS

Other pollution problems may arise from heavy metals, hot-water influx from power stations, radioactive wastes, pesticides and CFCs. Check with your syllabus to find out if you need to know about any of these. A knowledge of the effects of carbon dioxide, hydrocarbons and oxides of nitrogen in causing the greenhouse effect will also be of advantage. These gases form a screen reflecting infra-red rays back to earth from the atmosphere.

CONSERVATION

Biological conservation is concerned with the interaction between humans and the environment. It is, in the developed world, one of the main political, social and economic issues. This is because, in order for the environment to be conserved, there must be a reduction in the use of all animal, vegetable and mineral resources.

RESOURCE MANAGEMENT

Conservation broadly equates with the management of biological and physical resources and also with the regulation of the demands which are made by the human population. It may involve some destructive processes such as:

- culling deer to prevent the destruction of saplings in a wood
- burning gorse to rejuvenate heathland
- the total neglect of an area to allow a climax community to develop

Most often, biological conservation processes are concerned with maximising output from an area by avoiding either over- or under-exploitation.

Check with your own syllabus. You may need to know about the work of national or local conservation bodies. Some syllabuses expect you to have studied specific conservation activities at first hand.

PRACTICAL WORK

Most of the aspects of ecology can best be studied first hand and fieldwork is an essential part of your study. This work should include a long-term investigation of an ecosystem as well as short-term studies of populations and communities at a given time. Apart from being a study in its own right, ecological fieldwork is also a valuable adjunct to your work on classification, the variety of organisms, modes of life and life cycles.

You will need to become familiar with quantitative methods of investigating animal and plant populations and with the various techniques of recording and analysing the data you have obtained. These methods are covered in detail in textbooks of quantitative ecology and in some of the standard A-level biology texts.

Your syllabus may state a specific ecosystem or ecosystems which have to be studied, but if not, any ecosystem can be used to practise the techniques involved. To become thoroughly versed in ecological theory and practice it would be wise to carry out both an autecological and a synecological investigation.

Water pollution can be studied by abiotic means, such as finding BOD and other physicochemical characters of water, and also by studies of the presence and absence of indicator species. Air pollution studies are most easily carried out by lichen studies as these organisms are indicator species for levels of sulphur dioxide pollution.

Further reading
Alma, P J, 1993, *Environmental Concerns*. CUP
Cornwell A, 1983, *Man and the Environment*. CUP
Dowdeswell W H, 1984, *Ecology, Principles and Practice*. Heinemann
Green W P O, Stout G W and Taylor D J, 1984, *Biological Science*, 'Organisms and their environment'; Ch 13 'Quantitative ecology'. CUP
Smith R C, 1980, *Ecology and Field Biology* (3rd edn). Harper and Row

EXAMINATION QUESTIONS AND ANSWERS

QUESTIONS

Question 16.1
The diagram illustrates some of the relationships between plants and animals in an ecosystem:

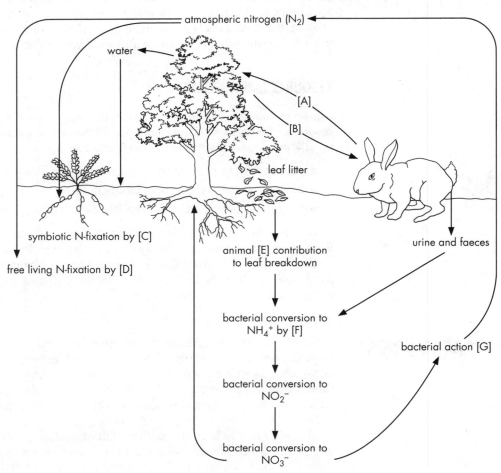

Fig. 16.9

a) Give a suitable name to each of the substances A and B. *(2)*
b) Give a suitable name to each of the organisms C, D and E. *(3)*
c) Give a suitable name to each of the processes F and G. *(2)*

(Total 7 marks)
(ODLE)

Question 16.2
The table shows the amount of nitrogen present in different forms in a freshwater lake at two times in the year.

AMOUNT OF NITROGEN/mg dm^{-3} PRESENT AS	AUGUST (SUMMER)	NOVEMBER (EARLY WINTER)
Organic nitrogen in living organisms	1.48	0.36
Ammonium ions	0.09	0.40
Nitrate ions	0.01	2.00
Nitrite ions	0.00	0.01

a) i) Calculate, the percentage of total nitrogen present as organic nitrogen in living organisms in August and in November. *(1)*
   ii) Suggest an explanation for the difference in these figures. *(2)*
b) High concentrations of nitrates can lead to vigorous growth of aquatic algae. Explain how this may lead to low levels of dissolved oxygen. *(3)*

(Total 6 marls)
(AEB)

Question 16.3
a) Describe what you understand by the following:
   i) pyramid of biomass
   ii) ecological climax *(8)*
b) What part do the following play in an ecosystem?
   i) saprophyte fungi
   ii) herbivores *(8)*
c) Why is it considered important to conserve ecosystems? *(4)*

(Total 20 marks)
(ULEAC)

Question 16.4
Give a *biological* explanation of why a farmer
a) plants a crop of clover after harvesting a cereal crop
b) should control the amount of nitrogenous material draining into a river
c) applies organic manure instead of inorganic fertilisers *(6)*

(Total 6 marks)
(AEB)

Question 16.5
An investigation was carried out into the effects of deep-burrowing earthworms and shallow-burrowing earthworms on the growth of roots in barley plants.

Four experimental plots of ground A, B, C and D, of equal size, were used in the investigation. Plots A, B and C were sterilised by fumigation (treatment with poisonous gas); plot D was left unfumigated. After fumigation, plot A was inoculated with deep-burrowing earthworms and plot B was inoculated with an equal number of shallow-burrowing earthworms. No earthworms were added to plots C and D. The treatments given to the four plots are summarised in the following table.

PLOT	STERILISED BY FUMIGATION	EARTHWORMS ADDED
A	Yes	Deep burrowing species
B	Yes	Shallow burrowing species
C	Yes	None
D	No	None

After these treatments each plot was sown with the same number of barley seeds. Seven months later the roots of the barley plants from each plot were examined. For each plot in turn all the root material found in the top 2 cm of soil was placed in one sample. Similarly, root material found in the next layer of soil, 2–4 cm deep, was placed in the next sample and so on in 2 cm layers to a depth of 8+ cm.

The root material per plant in each 2 cm layer of soil was determined and the results were shown in the graphs below.

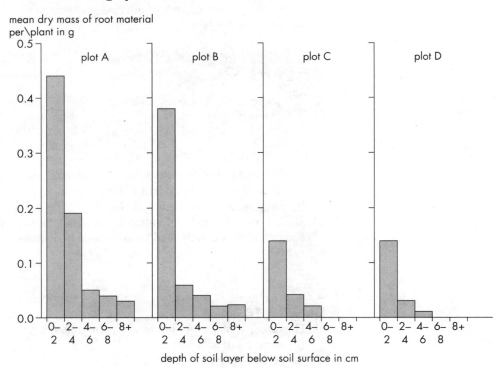

Fig. 16.10

a) i) What is the mean mass of a whole root of a plant from plot A? *(2)*
   ii) State **two** ways in which roots from plot C differ from roots from plot A. *(2)*
b) i) Compare the effects of the activities of deep-burrowing and shallow-burrowing earthworms on the growth of barley roots. *(3)*
   ii) Suggest **three** ways in which these effects may have been brought about. *(3)*
c) i) Explain the purpose of fumigating plots A, B and C. *(2)*
   ii) Explain why plot D was left unfumigated. *(2)*
d) Suggest ways in which a farmer could maintain high earthworm populations in the soil. *(3)*
e) Explain the ecological significance of the fact that different earthworm species live at different depths in the soil. *(3)*

(Total 20 marks)
(ULEAC)

OUTLINE ANSWERS

Answer 16.1
a) A = CO_2
   B = O_2
b) C = *Rhizobium* sp.
   D = Cyanobacteria or *Azobacter*
   E = earthworms or insects or myriapods, etc.
c) F = ammonification
   G = denitrification

Answer 16.2
a) i) August: 99.3%
      November: 13.0%
   ii) In August there is a large population of both plants and animals. The plants are growing quickly and assimilating inorganic nitrogen. In November most of the organisms have died and decayed, increasing the inorganic nitrogen present.

b) The increased plant growth gives rise to increased quantities of dead material. This is decomposed by aerobic bacteria. The bacteria increase in numbers rapidly and use up the available oxygen leading to low levels in the water.

Answer 16.3
a) i) Mass of organisms in each trophic level; measured as dry mass; at a given time; usually mass of producers at bottom largest and mass of top consumers smallest; does not indicate productivity.
   ii) State of an ecosystem in equilibrium; result of a succession of communities; population numbers fluctuate about the carrying capacity; type of climax community dependent on abiotic factors; usually woodland.
b) i) Heterotrophic; obtain organic matter from dead organisms; or from excreta and other wastes; secrete external enzymes; recycle plant nutrients.
   ii) Primary consumers in food chains; obtain organic compounds from autotrophs; prey for secondary consumers; loss of energy when autotrophs pass to consumers; can affect plant succession by eating saplings, etc.
c) Ecosystem preserves a diversity of organisms; provides a reservoir of genotypes; loss of one ecosystem affects another; may provide products not yet investigated – e.g. medicines; maintain gaseous equilibrium, especially rainforests.

Answer 16.4
a) Clover is leguminous plant with root nodules containing nitrogen-fixing bacteria; when the clover is ploughed in, the nitrogen content of the soil is increased.
b) Nitrogenous material adds excess nutrient to the water; accelerates the process of eutrophication; effect particularly severe if the river drains into a lake.
c) Organic manure increases the humus content of the soil; improves the soil structure; provides slow release of nutrients.

Answer 16.5
a) i) $0.44 + 0.19 + 0.05 + 0.04 + 0.03 = 0.75$ g
   ii) 1. roots in plot C are much shorter
       2. much less mass of root in plot C, less than $\frac{1}{3}$ that of plot A
b) i) Deep-burrowing earthworms allow the roots to grow longer; they increase the mass of root growth; deep-burrowing earthworms have a greater effect at all levels in the soil.
   ii) Increased soil aeration; increased soil drainage; increasing the depth of top soil by breaking up the subsoil.
c) i) Fumigation would kill the soil fauna; would remove the existing earthworm population and any herbivores which would damage the roots.
   ii) Plot D left unfumigated as a control; would give an indication of root growth in natural conditions.
d) Cut down on artificial (inorganic) fertilisers; use more organic fertilisers; leave the field fallow for some years.
e) Earthworms all have similar requirements for food; living at different depths means each species is in a different niche; reduces competition between the different species.

QUESTIONS, STUDENT ANSWERS AND EXAMINER COMMENTS

Question 16.6
a) Plants in a grassland ecosystem assimilate about 1% of the sun's radiant energy that falls on them. Approximately 80% of the plant's assimilated energy is available to the next trophic level.
   i) Give **two** reasons why plants assimilate so little of the sun's energy.

   1 Energy is reflected from the shiny cuticle of the leaf.
   2 Only some of the energy is of the correct wavelength to be used in photosynthesis

   (2)

> 66 Correct. Could also have said that some light passes through the leaf without being absorbed 99

ii) What happens to the energy assimilated by plants which is not available to the next trophic level?

Such energy is used by the plant itself: for respiration, synthesis, growth. It is alternately lost as heat. (2)

> **Use in respiration is correct but that used in growth (synthesis) forms biomass which can be passed on to the next level**

b) Explain why the number of links in food chains does not often exceed five.

Only a small amount of energy at any trophic level is available to the next. This is stored energy, the rest being used by the organism. (6)

(Total 10 marks)
(AEB)

= excretion, respiration, heat — Each further trophic level can obtain less and less energy from the level below. After 5 links the stored energy is usually insufficient to support another organism.

> **The points are made that only a small proportion of the energy is passed on and there is insufficient energy available to support more than five levels. The answer is more wordy than necessary. Note that it is much longer than the space provided**

Question 16.7

The following data relate to two separate species of leaf-mining moths. The larvae of one species mine in the leaves of birch trees and the larvae of the other species mine the leaves of oak trees. The data were obtained during an investigation of mortality factors affecting the two species of moths.

STAGE IN LIFE HISTORY	NO. OF INDIVIDUALS SURVIVING TO THE END OF EACH STAGE AS A PERCENTAGE OF THE NUMBER ENTERING THE STAGE.	
	Birch leaf miner	Oak leaf miner
Instar 1	92	95
2	95	90
3	82	73
4	65	70
5	45	64
Prepupa	66	71
Pupa	80	64

a) i) If 100 eggs were hatched, calculate the actual number of beech leaf miners which could be expected to survive to the end of the fifth instar stage. Clearly show your method of calculation and correct to the nearest whole number at each stage of your calculation.

1st $\quad \frac{92}{100} \times 100 = 92$

2nd $\quad \frac{95}{100} \times 92 = 87$

> **Calculations correct**

3rd $\quad \frac{82}{100} \times 87 = 71$

4th $\quad \frac{65}{100} \times 71 = 46$

> **Correct**

5th $\quad \frac{45}{100} \times 46 = 21$

ii) Given that the survival rate of the comparable fifth instar of the oak leaf miner is 28%, comment on the survival rates of the two species.

Both have variable survival rates at different stages. The survival rates of 28% and 21% at the fifth instar are similar. (6)

> **Incorrect, the survival rates at the fifth instar are significantly different**

b) State **two** ways in which the survival pattern to the end of the pupal stage of the oak leaf miner compares with that of the birch leaf miner.

> **Correct**

　i)　Greater survival overall by oak miners

　ii)　Birch miners survive better in the pupal stage.　　　　　　(2)

c) Suggest **four** factors which could limit the population of leaf-mining species during their life cycles.

　i)　Loss of leaves from the trees

> **Correct**

　ii)　Being parasitised

　iii)　Predation

　iv)　Temperature　　　　　　(4)

> **Insufficient, should state temperature too low (or too high)**

(Total 12 marks)
(ULEAC)

REVIEW SHEET

1. Define each of the following, providing examples wherever appropriate.
   a) Ecosystem: _____
   _____
   b) Population: _____
   _____
   c) Habitat: _____
   _____
   d) Ecological niche: _____
   _____

2. a) Complete the boxes in the following diagram of the **nitrogen cycle**.

   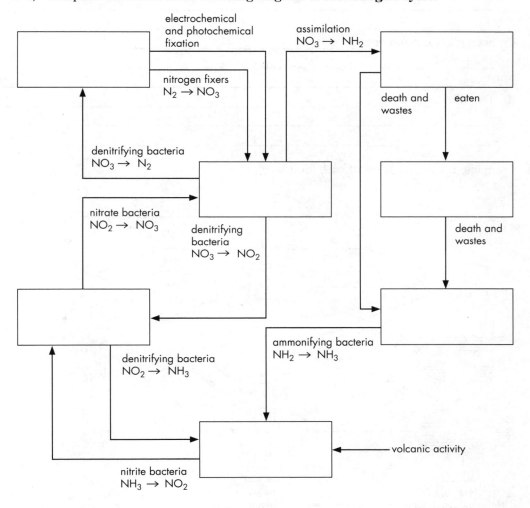

   b) Now use the diagram to explain each of the following processes within the nitrogen cycle.
   Putrefaction _____
   _____
   _____
   Nitrification _____
   _____
   _____

Nitrogen fixation _____
_____
_____

Denitrification _____
_____
_____
_____

3 Define each of the following terms used in connection with food chains and webs, giving examples where appropriate.

a) Producers: _____
_____
_____

b) Primary consumers: _____
_____
_____

c) Secondary consumers: _____
_____
_____

d) Trophic level: _____
_____
_____

4 Complete the following **food web** for a pond.

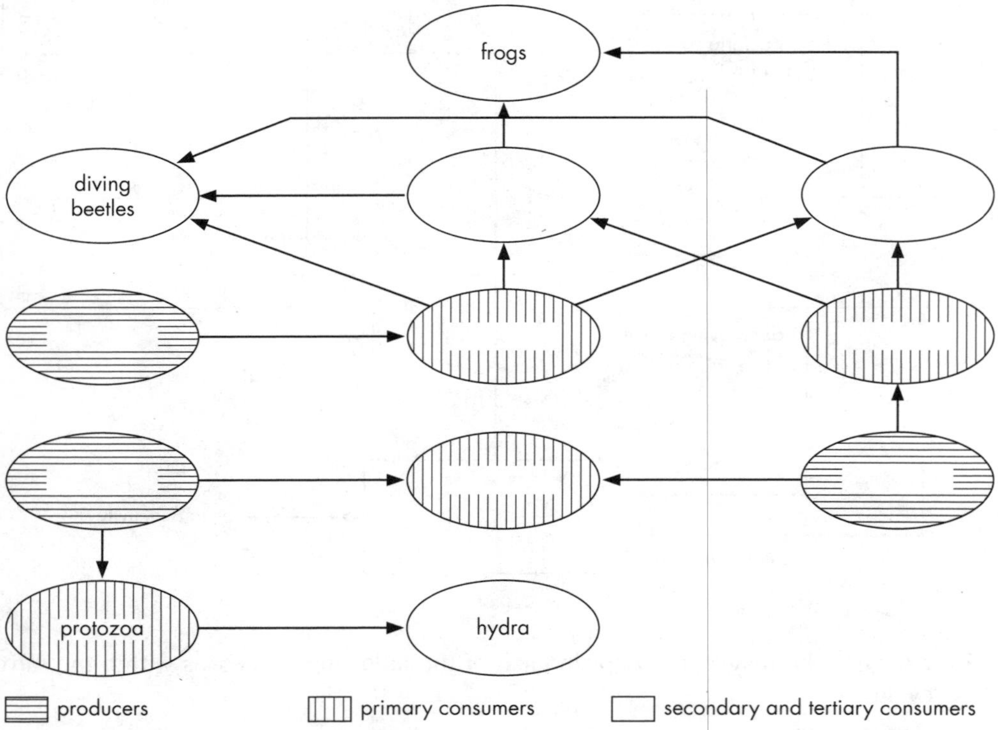

5 Consider each of these three **pyramids of numbers**, then suggest what each represents.

A: _____

B: _____

C: _____

6 Consider each of these two **pyramids of biomass**, then suggest what each represents.

A: _____

B: _____

7 Briefly outcome some of the factors involved in determining the rate of **population growth.**

_____
_____
_____
_____
_____
_____

8 Define each of the following, giving examples where appropriate.
  a) Autecology: _____
  _____
  _____
  _____

  b) Synecology: _____
  _____
  _____
  _____

CHAPTER 17

BIOTECHNOLOGY AND GENETIC ENGINEERING

FERMENTATION

GENETIC ENGINEERING

ENZYMES

IMMUNOLOGY

HUMAN DNA MANIPULATION

FUTURE TRENDS

PRACTICAL WORK

GETTING STARTED

Biotechnology is the application of biological processes or systems to manufacturing and service industries. The agents used may be micro-organisms, single cells or biological substances such as enzymes.

Genetic engineering involves the **in vitro** manipulation of DNA. It involves the introduction of manipulated DNA into cells in such a way that it will replicate and be passed on to progeny cells. DNA from various organisms is introduced into bacterial cells, which then produce the required product. The techniques developed in this work are also being used in the clinical field with research into changing human DNA (gene therapy), chromosome sequencing and in the forensic field (DNA finger printing).

A-level syllabuses do not expect a very detailed knowledge of the techniques involved but they require an understanding of their basics and an appreciation of present and future applications of both biotechnology and genetic engineering.

CHAPTER 17 BIOTECHNOLOGY AND GENETIC ENGINEERING

ESSENTIAL PRINCIPLES

FERMENTATION

" Note these two definitions of fermentation "

Although it is old technology, fermentation is the most important area of biotechnology, with a worldwide turnover of about £10 billion.

- **Biochemically**, fermentation is the catabolic breakdown of substrates in anaerobic conditions.
- **Industrially**, however, the term is used to describe the production of commodities by micro-organisms even if the processes are aerobic.

PRODUCTS OF FERMENTATION

The main products fall into four categories: microbial biomass; microbial metabolites; microbial enzymes; transformation process products.

Microbial mass

Microbial mass is produced on a large scale by continuous culture techniques and is an important source of commercial protein. The product, known as **single cell protein**, is used as animal feed and research is taking place to produce human food from fungal biomass.

Microbial metabolites

Microbial metabolites include many important commercial products:

- **ethanol** for beverages and industrial use is produced by *Saccharomyces cerevisiae*
- **citric acid** for use in the food industry comes from the culture of *Aspergillus niger*
- **glutamic acid** and **lysine**, food additives, are produced by *Corynebacterium glutamicum*
- **acetone** and **butanol,** the industrial solvents, are produced by *Clostridium acetobutyricum*

All these products are primary metabolites. Some micro-organisms also produce secondary metabolites from primary metabolites and many of these are commercially useful. For example:

- various antibiotics
- some cancer treatment drugs (e.g. Bestatin)
- immuno-suppressants such as Cyclosporin A
- plant-growth regulators (gibberellin)

Microbial enzymes

Microbial enzymes are produced for use in the food industries, for use in biological washing powders and for clinical and analytical purposes. As well as producing microbial enzymes it is now possible to engineer microbial cells to produce eukaryote enzymes: for example, the production of **rennin** by *Escherichia coli*.

Transformation process products

Transformation processes are where microbial products are converted into other products. The oldest example is the production of vinegar from ethanol. Many different products can be produced by the catalysis of many different biochemical reactions, such as oxidation, dehydrogenation, deamination, amination and isomerisation.

GENETIC ENGINEERING

Genetic engineering or **gene manipulation** has opened new doors in the field of selective breeding and has also made it possible to produce biochemical agents from eukaryotes in large quantities by putting eukaryote genes into prokaryote cells. In time the manipulation of genomes in higher animals and plants will be available but at present genetic engineering is really effective only in single-celled organisms.

GENETIC INFORMATION TRANSFER

Genetic information can be transferred from one micro-organism to another by vector particles. These take two forms:

- **plasmids**, small circular loops of DNA found in many bacterial cells
- **phages** or bacterial viruses

Plasmids
Plasmids replicate independently from the host cell and code for functions such as antibiotic resistance. Plasmids can pass from one cell to another between closely related species when conjugation takes places.

Phages
Infection by phages is also limited to just some groups of bacteria. The techniques of genetic engineering overcome these limitations.

GENE MANIPULATION

There are four stages in gene manipulation:

- the formation of DNA fragments including the gene wanted for replication
- the splicing (insertion) of the DNA fragments into the vector
- the introduction of the vector into the bacterium
- the selection of the newly transformed organisms for cultivation

FORMATION OF DNA FRAGMENTS

DNA fragments are produced by using one of the **restriction enzymes**. Each restriction enzyme recognises a specific base sequence and splits the DNA molecule within that sequence. Most restriction enzymes split the two strands in a staggered sequence, forming 'sticky ends', as shown in Fig. 17.1.

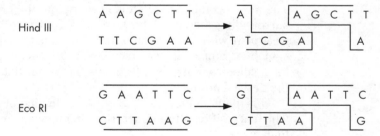

Fig. 17.1 Action of two restriction endonucleases: Hind III, Eco RI

Longer eukaryote fragments can be produced from purified mRNA using the enzyme **reverse transcriptase** to produce a single strand of DNA and **DNA polymerase I** to form the partner strand. The 'sticky ends' are then added to the manufactured strand.

SPLICING

The DNA fragments containing the gene for the desired product are inserted into a vector. A manufactured plasmid, pBR322, is most used, as in the transformation *E. coli*. This plasmid carries genes for resistance to tetracycline and ampicillin and so can be selected by the use of these antibiotics in the growth media. It is a small plasmid and has unique cleavage sites for several restriction enzymes. The steps to insert a gene into a plasmid are shown in Fig. 17.2. The plasmid and the required DNA are treated with the same restriction enzyme – for example, Hind III, giving an open plasmid and DNA fragments with complementary 'sticky ends' which are joined by mixing and adding the enzyme **DNA ligase**.

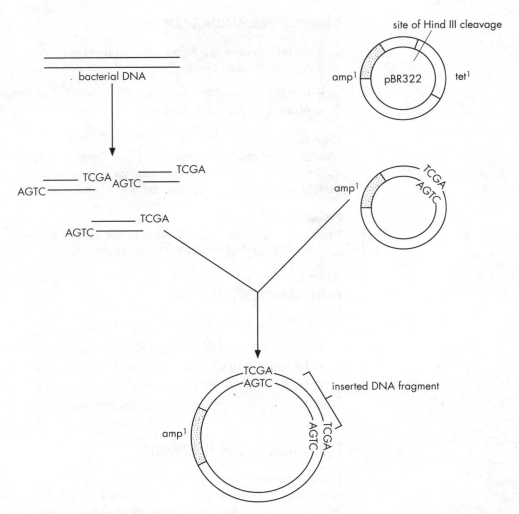

Fig. 17.2 Insertion of DNA into a plasmid vector

TRANSFORMATION AND SELECTION

The new recombinant DNA is now put into the host bacterial cell by uptake from the surrounding medium. The host cells are made 'competent' to take up the DNA by treatment with calcium chloride. The transferred bacteria are plated out on to agar containing ampicillin or tetracyline. Those containing the plasmid are resistant and will form colonies. Further selection will sort out the bacteria which contain the gene for the desired product.

This methodology has led to the production of **hormones** and other **polypeptides** which are difficult to extract from mammalian sources. The most important products to date are:

- somatostatin
- insulin
- interferon

ENZYMES

An **enzyme** is a complex three-dimensional globular protein which acts as a catalyst in living systems – that is, a substance which, in small amounts, promotes chemical change without itself being used up or altered in the reaction. The use of enzymes for cheese-making and hydrolysing starch to sugar are again ancient forms of biotechnology and enzyme technology is now an important aspect of the science.

PURIFICATION

Most enzymes used in industry are derived from micro-organisms. Some enzymes are extracellular but others have to be released from cells. Once freed the enzymes have to be purified.

- Solid material can be removed by centrifuging or filtering.
- Nucleic acids are precipitated by specific agents or hydrolysed by nucleases.
- The proteins are then precipitated out by adding ammonium sulphate to the solution as proteins are less soluble in high-ionic-strength solutions.
- The required enzyme is then concentrated and purified by techniques such as chromatography.

FORMS OF USABLE ENZYME

When the recovery of the enzyme after use is unimportant, as in biological detergents, the enzyme is used in soluble form. If the recovery and re-use of the enzyme is important, the enzyme is immobilised by combination with an insoluble chemically inert material such as cellulose. The enzyme may be chemically bound to a polymer or enclosed within a membrane or gel.

INDUSTRIAL APPLICATIONS

Detergents
About a third of the enzymes produced are used in detergents. Biological washing powders contain proteases, mainly **subtilisin** from *Bacillus subtilis*, which dissolves protein-containing material.

Starch and sugar processing
Several enzymes are used in starch and sugar processing, mainly amylases and **glucose isomerase**, which converts glucose to fructose.

Other uses
Enzymes are also used in:

- amino acid production
- the removal of hairs from hides in leather making
- the removal of starch from textiles

BIOSENSORS

These are a combination of an immobilised enzyme and an electrode. The electrode detects changes in substrate or product, temperature changes or optical properties. They are used in:

- hospital and food-science laboratories for monitoring toxic compounds in environmental control
- industrial processes such as fermentation control

IMMUNOLOGY

MONOCLONAL ANTIBODIES

Antibodies are very useful in the diagnosis and treatment of disease, for purifying small amounts of rare compounds and for detecting and quantifying chemicals in the body. The small amounts present and the structural similarity of different antibodies has made it very difficult to purify single antibodies, added to which antibody-producing cells fail to grow in tissue culture. The problem was solved by the production of monoclonal antibodies. Antibody-secreting cells are fused with myeloma (tumour) cells to form **hybridomas**. These hybrid cells will grow indefinitely in tissue culture and secrete antibody (see Fig. 17.3). Hybridomas producing different antibodies can be isolated and grown in bulk to form clones.

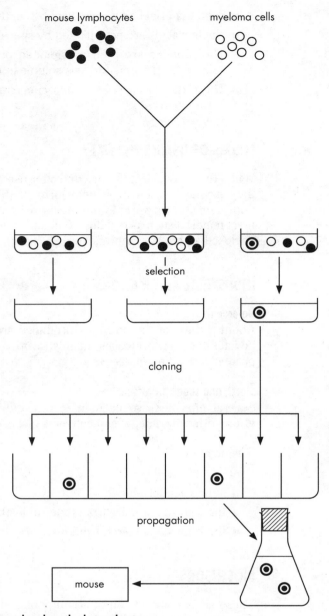

Fig. 17.3 Principle of monoclonal antibody production

Example of monoclonal antibody production
The technique shown in Fig. 17.3 is as follows.

- Spleen cells from an immunised mouse are the source of lymphocytes, some of which produce the required antibody.
- They are fused with myeloma cells and grown in tissue culture with a selecting agent so that only the fused cells survive.
- These are then recultured at very low densities so that clones are produced from individual cells.
- The new cultures thus produce pure antibody.

Uses of monoclonal antibodies
Monoclonal antibodies are now used for many diagnostic purposes, such as assaying alpha fetoprotein levels in amniotic fluid to detect spina bifida. In time the technique will also be useful for targeting drugs to specific cells in the body for therapeutic purposes.

HUMAN DNA MANIPULATION

Recombinant DNA technology is now being used in diagnostic and forensic medicine. Eventually it may be useful in both preventative and therapeutic medicine.

DIAGNOSIS OF GENETIC DISORDERS

Procedure

- DNA is isolated from the patient's white blood cells and fragmented using sequence-specific restriction endonucleases.
- The fragments are separated by electrophoresis, then transferred to a filter.
- The filter is then incubated with a radioactively labelled copy of a specific gene, a **gene probe**, which combines only with its complementary sequence on the filter.
- The DNA fragments containing some or all of the probe sequences are then found by using autoradiography.

This technique is the basis for the diagnosis of many genetic disorders.

MAPPING A GENETIC DISEASE

There are some neutral mutations within or close to genes which alter the cleavage pattern of a gene with a particular restriction enzyme. These mutations are called **restriction enzyme polymorphisms (REP)**. Hundreds of REPs from different chromosomal regions are now known and so tests can be carried out until one is found which is closely linked to the affected gene under consideration. This technique has been used to find the genes responsible for many important inherited defects, such as:

- cystic fibrosis
- sickle cell anaemia
- Huntington's chorea

DNA FINGERPRINTING

In human chromosomes there are **hypervariable regions (HVR)** made up of tandem repeats of a short nucleotide sequence, 15 to 30 base pairs in length. They are hypervariable because the number of tandem repeats and consequently also the length of the DNA in that region varies greatly in the population. DNA probes have been produced which detect these HVRs at many different loci. The probes developed detect 30 to 40 different HRVs at once and so the chances of two individuals having the same length is infinitesimally small. The banding pattern obtained in the autoradiograph spots (Southern blots) of human DNA hybridised with these probes is therefore specific to individuals and is called a **genetic fingerprint**. The bands in a fingerprint show Mendelian inheritance and have already been used to resolve questions of paternity and to convict rapists.

> **FUTURE TRENDS**
>
> 66 Read this section carefully. You must distinguish between what is possible at present and what may be possible in future 99

So far only a start has been made in the development of genetic engineering and its application to biotechnology. There are several fields in which research is taking place and although progress is slow, major advances will be made over the next few decades.

HEALTH CARE

Therapeutic products

New therapeutic products will be developed, including human hormones and new vaccines. Several genetically engineered vaccines are already under development, including one against malaria.

Gene therapy

Gene therapy will become available where cells containing non-mutant genes will be substituted for the abnormal mutant cells. At present, work is taking place to replace abnormal bone marrow cells with normal ones to correct illnesses such as β thalassaemia caused by genes for abnormal globins. This is an area in which a breakthrough would allow the treatment or prevention of many inherited illnesses.

CHAPTER 17 BIOTECHNOLOGY AND GENETIC ENGINEERING

Implantation
Further into the future is the possibility of implanting genetically engineered cells to produce insulin in diabetics or liver cells to treat liver failure.

AGRICULTURE

Hormones and vaccines
Genetically engineered growth hormones are already available to increase meat yield and work is in progress to produce cows giving milk more suitable for conversion to long-life milk. There is also work in progress to engineer vaccines against foot and mouth disease, trypanosomiasis and other animal diseases.

Genetic manipulation of crops
There is great potential for improving crop plants by genetic manipulation. One of the aims is to produce cereals which can fix nitrogen and another is to improve cereal proteins so they contain more essential amino acids.

INDUSTRY

Tailor-made enzymes
The use of microbes and enzymes in industry will increase and new enzymes will become available. Genetic manipulation will be used to improve enzymes, to increase their turnover or to improve their temperature stability and pH stability. Such 'tailor-made' enzymes could produced for specific industrial processes which are at present inefficient.

PRACTICAL WORK

Practical work can be carried out investigating various aspects of fermentation by microbes and assaying the products in different conditions. In this context, fermentation means any production of metabolites by micro-organisms, not just the study of anerobic processes.

There are now mobile biotechnology centres and a visit to one of these would be advantageous to obtain more information about the techniques used in both research and industry.

Further reading
Lowrie P, and Wells S, 1991, *Microorganisms, Biotechnology and Disease*. CUP
Smith J E, 1981, *Biotechnology*. Arnold
Walker J M and Gingold E B (eds), 1988, *Molecular Biology and Biotechnology (2nd edn)*. Royal Society of Chemistry
Warr J R, 1984, *Genetic Engineering in Higher Organisms*. Arnold.
Wiseman A (ed), 1983, *Principles of Biotechnology*. Surrey University Press

EXAMINATION QUESTIONS AND ANSWERS

QUESTIONS

Question 17.1
Suggest how genetic engineering may be applied to the production of enzymes. *(4)*
(Total 4 marks)
(ULEAC)

Question 17.2
Read the following passage on genetic engineering and write on the dotted lines the most appropriate word or words to complete the passage.

The isolation of specific genes during a genetic engineering process involves forming eukaryotic DNA fragments. These fragments are formed using .... enzymes which make staggered cuts in the DNA within specific base sequences. This leaves single-stranded 'sticky ends' at each end. The same enzyme is used to open up a circular loop of bacterial DNA which acts as a .... for the eukaryotic DNA. The complementary sticky ends of the bacterial DNA are joined to the DNA fragment using another enzyme called .... DNA fragments can also be made from .... template. Reverse transcriptase is used to produce a single strand of DNA and the enzyme .... , catalyses the formation of a double helix. Finally new DNA is introduced into host .... cells. These can then be cloned on an industrial scale and large amounts of protein harvested. An example of a protein currently manufactured using this technique is. ....

(Total 7 marks)
(ULEAC)

Question 17.3
Read through the following passage and then answer the questions.

1   Genetic engineering is not new: the importance of DNA for all living processes has long been appreciated. It provides the blueprints for life, not only determining by a precise and reproducible arrangement of purine and pyrimidine bases in a double helix, the heredity of the cell, be it of bacterial or
5   human origin, but at the same time coding for every function of the cell including the deployment of enzyme systems which will allow the DNA double-helix to repair itself if it becomes damaged. As early as 1967 an enzyme, DNA-ligase, had been discovered which joined the breaks in DNA chains. In 1972 some workers were making artificial genes – chains of DNA constructed
10 entirely in the test-tube and suitably, and specifically, joined together. The phenomenon of 'restriction' had been known earlier. Although certain small viruses could invade and multiply in certain bacteria, others could not. This 'foreign' DNA was, in fact, cut into pieces by bacterial enzymes called restriction endonucleases, and so rendered non-infective. A score or so of
15 different bacterial restriction enzymes have now been identified. DNA strands cannot only be chopped up into smaller pieces but methods are available for joining the ends together again, either to each other or to strange DNA. If the 'new' DNA was to be produced in quantity, then suitable DNA vehicles, or carriers, were required.

Adapted from 'Commonsense in Genetic Engineering' by R. Harris in Biologist (1977)

a) What is meant by 'a precise and reproducible arrangement of purine and pyrimidine bases' (line 3)? *(4)*
b) Indicate how DNA is involved in 'coding for every function of the cell' (line 5). *(5)*
c) Suggest how enzymes such as 'DNA ligase' (line 7) and 'restriction endonucleases' (line 13) could be used in genetic engineering. *(3)*
d) Describe *in outline* how 'viruses could invade and multiply in' cells (line 11). *(4)*
e) Suggest **one** potential danger and **one** potential benefit of genetic engineering to humans. *(4)*

(Total 20 marks)
(ULEAC)

Question 17.4
In many biotechnological processes, living micro-organisms are grown in nutrient solutions in fermenters like the one shown below. Useful substances produced by the micro-organisms can be extracted for industrial or other use.

The nutrient solution is cooled by the water jacket, aerated through the air line and mixed by stirring. Acids or alkalis may be added from the acid/alkali reservoir and steam lines are present to that the apparatus can be sterilised after each fermentation batch.

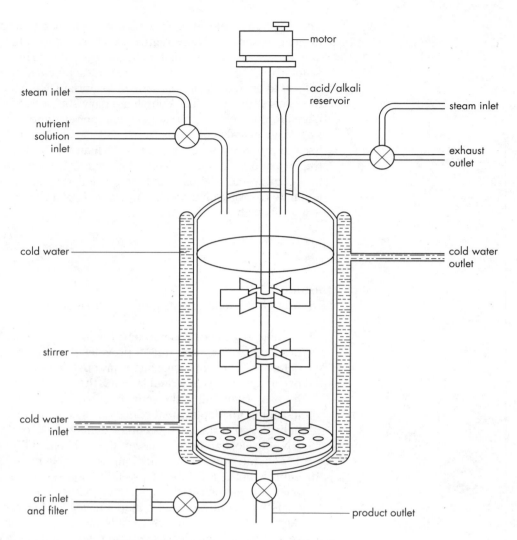

Fig. 17.4

a) Suggest **two** different types of micro-organism that can be grown in a fermenter like the one in Fig. 17.4. (2)
b) Explain the necessity for each of the following when operating such a fermenter:
   i) cooling
   ii) aeration
   iii) addition of acids or alkalis (6)
c) Using this type of fermenter, an experiment was carried out to investigate the effect of different carbon sources on the secretion of the enzyme amylase by a micro-organism.

   Five cultures of the same micro-organism were set up, each with only one carbon compound in its nutrient solution after 72 hours.

   The results are shown in the table below.

CARBON SOURCE IN MEDIUM	CONCENTRATION OF AMYLASE AFTER 72 HOURS IN UNITS cm^{-3}
Starch	235
Maltose	179
Glucose	52
Sucrose	17
Lactose	3

i) Suggest **two** other nutrients, excluding water, that you would expect to be provided in the nutrient solution in this experiment. (2)
ii) Name the type of nutrition shown by the micro-organism in this experiment. (1)
iii) State **four** precautions that should be taken in carrying out this experiment to ensure that the results for the different cultures are comparable. (4)

iv) Compare and comment on the effects of the various carbon sources on amylase production in this micro-organism. *(5)*
(Total 20 marks)
(ULEAC)

OUTLINE ANSWERS

Answer 17.1
Identification of the gene for the particular enzyme; produce the relevant DNA fragment; insert fragment into vector; introduce vector into host bacterial cells; host cells produce the required enzyme.

Answer 17.2
Endonuclease; vector; DNA ligase; mRNA; DNA polymerase; bacterial; insulin/growth hormone/interferon/etc.

Answer 17.3
a) DNA contains a chain of purine and pyrimidine bases; each gene has a specific sequence of bases which is always the same; when DNA replicates the chain of bases is copied exactly; because each base can pair only with a particular partner, that is A with T or C with G.
b) Each gene is a particular piece of DNA; each produces a specific polypeptide; each polypeptide forms a specific protein or enzyme; these bring about the metabolic activity in the cell; the function of the cell is controlled by the coding in the DNA.
c) Restriction endonucleases can be used to produce fragments of mammalian DNA; this can then be put into the DNA of bacterial plasmids or viruses; it is fastened into the bacterial DNA by the DNA ligase.
d) Virus attaches to cell membrane; viral nucleic acid enters cytoplasm of the cell; causes the cell to replicate it; a new protein envelope is formed for each piece of viral nucleic acid so new viruses are formed.
e) Danger; pathogenic organisms may be accidentally produced; there may be no drugs available to deal with them. Benefit: manufacturing hormones, for example insulin or somatotrophin; by placing human DNA into a bacterial plasmid.

Answer 17.4
a) Bacteria
   Fungi
b) i) Fermentation produces heat; cooling prevents temperature rising above the optimum.
   ii) Aerobic respiration favoured; more energy released for faster growth.
   iii) Substances secreted by the organisms may alter pH; may be slow growth if pH not optimum.
c) i) Nitrates; phosphates
   ii) Heterotrophic
   iii) Same temperatures; equal masses of carbon sources; equal volume of culture solution; equal amount of stirring.
   iv) High amylase production in presence of starch; starch is substrate for amylase; maltose is product of amylase action; low amylase production in presence of sucrose or lactose; these sugars may inhibit amylase synthesis.

QUESTION, STUDENT ANSWER AND EXAMINER COMMENTS

Question 17.5
Read the passage below and then answer the questions which follow.

Insulin from pigs and horses has been widely used for treating diabetes mellitus, but it is not identical to human insulin and can have undesirable side effects. Human insulin can now be produced commercially using the techniques of genetic engineering. Reverse transcriptase may be used to generate cDNA from intron-free insulin mRNA. 'Start and stop' signals are added to the cDNA, together with sticky ends. The gene is then inserted into a suitable plasmid, such as pBR322, and the hybrid vectors are mixed with bacteria in conditions which encourage their uptake. The latter are then cloned, using suitable enrichment techniques, and screened for insulin production using DNA probes or antibodies. Finally the required clone is cultured, and the insulin is extracted and purified. One problem often found is that recombinant bacteria may lose their desired hybrid plasmids. In the absence of selective media the resulting daughter cells grow at a faster rate than their hybrid parents, so the rate of insulin production in a culture may decrease.

a) In what way is reverse transcriptase an unusual enzyme?

> It makes DNA from RNA   **" Correct "**

(1)

b) What is an intron?

" Correct but not enough for two marks. Should go on – which does not code for amino acids "

> A nucleotide sequence within a gene.

(2)

c) Human insulin genes, which contain two small introns, can be isolated from human chromosomes, but even if correctly incorporated into bacteria they cannot produce insulin. Why not?

" Should go on to say due to altered coding "

> Bacteria do not have introns and cannot process introns and so they produce abnormal protein from the introduced gene.

(3)

d) What is a clone?

> Genetically indentical individual.   **" Correct "**

(1)

e) What is meant by the term hybrid vector?

" Should go on to say used to incorporate this DNA into the bacterium "

> A plasmid or phage with recombinant DNA.

(2)

Fig. 17.5

f) A plasmid vector often used for genetic engineering is pBR322, shown below:

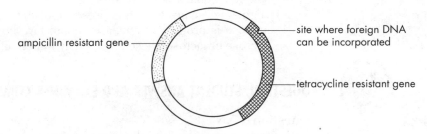

pBR322 was mixed with insulin cDNA and subsequently with bacteria. The bacteria were afterwards grown on agar. The colonies were then replicated on to the media shown below.

GROWTH MEDIUM	+ AMPICILLIN ADDED	+ TETRACYCLINE ADDED
Type A	Grows	Grows
Type B	Grows	No growth
Type C	No growth	No growth

Three types of colonies were found (A, B and C), some growing on only one medium, some on neither and some on both. Give an explanation of these results and show how this procedure can be used to identify those colonies which contain hybrid vectors.

A good answer

```
Any media with antibiotic kills bacteria with no plasmids
because they have no resistance to antibiotic. Type C does
not contain any plasmids. Type B must contain hybrid
vectors as the inclusion of foreign DNA destroys the
tetracycline resistance.
```

(4)

g) Suggest why bacteria which have lost their hybrid plasmids grow faster than those which still possess them.

Enough for one mark, should also mention metabolites not used to produce plasmids

```
If there are plasmids in the cell, energy will be used to
produce more plasmids.
```

(2)
(Total 15 marks)
(ODLE)

NB Review questions to check your understanding of this chapter can be found at the end of Chapter 18.

CHAPTER 18
MICROBIOLOGY

- DIVERSITY OF MICRO-ORGANISMS
- NUTRITION
- CULTURE TECHNIQUES
- APPLICATIONS IN SOCIETY
- PRACTICAL WORK

GETTING STARTED

An increasing number of syllabuses, both for A- and AS-levels, are incorporating options or modules in microbiology. Even where no special or separate option exists, there are references in nearly all the syllabuses to the culturing of micro-organisms and their use in food production, the purification of water and the treatment of sewage, together with some appreciation of the principles and application of **genetic engineering**. This chapter outlines some of the areas of study which are included in recent syllabuses, but cannot possibly cover all the material needed in the required depth, so reference should be made to the relevant chapters in this book, particularly those dealing with the diversity of organisms, respiration, nutrition and biotechnology.

Much emphasis is placed on the practical work associated with culturing micro-organisms in the laboratory, and students following a course which involves such experiments should be aware of the hazards associated with the handling of micro-organisms. Strict aseptic techniques should be observed at all times. Some of the practical work referred to in this chapter must be carried out only in a suitably equipped laboratory and the waste material disposed of in a safe way by sterilisation.

CHAPTER 18 MICROBIOLOGY

ESSENTIAL PRINCIPLES

DIVERSITY OF MICRO-ORGANISMS

It is important to study a range of micro-organisms in order to appreciate their diversity with respect to size, complexity and reproduction. Most syllabuses will include representatives of both prokaryotic and eukaryotic groups.

KINGDOM PROKARYOTAE

From the Kingdom Prokaryotae, bacteria such as *Escherichia* and *Staphylococcus* as non-photosynthetic types, and *Anabaena* as a photosynthetic type, have been selected as suitable examples for the study of structure and reproduction. The use of Gram-staining as a method of subdividing the bacteria into two groups, the **Gram positive** and the **Gram negative**, is a technique which candidates are expected to know.

KINGDOM PROTOCTISTA

The Kingdom Protoctista contains a large number of micro-organisms and representatives from the following phyla are commonly referred to in the syllabuses:

- Rhizopoda – *Amoeba*
- Ciliophora – *Paramecium*
- Oomycota – *Saprolegnia*
- Chlorophyta – *Chlorella*

KINGDOM FUNGI

In the Kingdom Fungi, representatives from all the phyla need to be studied; many of the organisms referred to are of significant economic importance.

VIRUSES

Reference is also made to viruses, which are usually grouped according to their **nucleic acid type**. The examples which are most commonly given are **influenza** and **tobacco mosaic** viruses and the **bacteriophage T4**.
The study of the different types of micro-organisms should enable students:

- to recognise the major characteristics of the groups
- to identify photographs and drawings

NUTRITION

In order to be able to appreciate how micro-organisms can be made use of, it is necessary to understand their nutritional requirements. Usually the organisms for study have been selected to cover the range of different types of nutrition. A few of the selected micro-organisms are **autotrophic**, but the majority are **heterotrophic**.

AUTOTROPHIC MICRO-ORGANISMS

Amongst the autotrophic micro-organisms, members of the **Chlorophyta** and the **Cyanobacteria** photosynthesise in the same way as green plants, fixing carbon dioxide using light energy. Both groups contain photosynthetic pigments, which absorb the light energy.

Photoautotrophic bacteria
Photoautotrophic bacteria can photosynthesise in a similar way, but possess **green** or **purple bacteriochlorophyll** as their photosynthetic pigment and use **hydrogen sulphide** or **organic compounds** as their source of electrons instead of water. These bacteria do not release oxygen and are anaerobic.

Chemoautotrophic bacteria

Chemoautotrophic bacteria use **carbon dioxide** and other simple inorganic molecules to build up organic materials. The energy necessary for this is derived from oxidation or reduction reactions. Bacteria in this group play a very important role in the recycling of essential nutrient ions, such as nitrogen, iron and sulphur. Reference to the nitrogen cycle bacteria *Nitrobacter* and *Nitrosomonas* is often made.

HETEROTROPHIC BACTERIA

Heterotrophic bacteria need complex organic compounds to be supplied to them. Precisely which organic compounds varies from species to species; some bacteria can use simple sugars and mineral salts while others require specific vitamins and amino acids supplied in their environment before they will grow. Variations in nutritional requirements can be used to distinguish between different species.

FUNGI

The fungi are heterotrophic and use complex organic substances as a source of energy. They need an organic source of carbon, a source of nitrogen, mineral elements, trace elements and vitamins. Many different sources of carbon can be used, from simple sugars to complex carbohydrates such as cellulose and lignin, though only relatively few Basidiomycota can utilise the latter. Inorganic nitrogen compounds such as nitrates and ammonium salts can be used by some fungi, though others may require amino acids to be present.

CULTURE TECHNIQUES

This section of the topic consists almost entirely of practical work. Students are expected to have:

- prepared different types of sterile media
- prepared agar plates
- prepared liquid cultures for the culture of fungi, algae and bacteria as appropriate

In addition, there are specific areas students should be familiar with.

- Different methods of inoculation of media are also included in most syllabuses.
- The uses of fermenters and the principles of continuous culture of micro-organisms on a large (industrial) scale should be understood.
- The techniques for isolating micro-organisms from soil and water should be known.
- The techniques for isolating pathogenic organisms from infected tissues should be known.
- Organisms grown in culture can be used to measure growth rates by means of cell counts, to estimate dry weight and to measure turbidity.
- The technique of dilution plating should be understood.
- It is also important to realise that external factors such as the pH of the medium and the temperature will have an effect on the growth rate of micro-organisms, as with other organisms and reactions already studied in other sections of the syllabus.

It is not possible to go into the details of all the different techniques used to culture micro-organisms here, and you are recommended to use the references given at the end of the chapter.

APPLICATIONS IN SOCIETY

Micro-organisms can be used in a number of ways.

FOOD PRODUCTION

Man has been using micro-organisms in the production of food for centuries, long before anyone thought of the term biotechnology!

Yeasts

Yeasts are used in baking to give the bread dough a spongy texture, and in brewing to produce alcohol. Both of these processes exploit the respiratory activities of the yeasts.

Cheese and yoghurt

Cheese and yoghurt are produced by introducing bacteria into milk. These bacteria, *Lactobacillus* and *Streptococcus* spp, respire anaerobically and fermentation occurs. Lactic acid is produced, creating acid conditions, which prevent the growth of harmful bacteria. Acid conditions also cause the protein in the milk to solidify. Under these conditions yoghurt will keep for up to three weeks, whereas the untreated milk would go bad in a few days. In cheese production the solid protein is precipitated and water removed from the curds.

Single-cell protein

In response to the demand for food both for humans and for livestock, micro-organisms are grown for their protein. They contain proteins and oils and it is possible to grow them on poor or waste materials. The cells can be harvested, the protein and lipid extracted, purified and processed. When this technique is used to produce protein it is referred to as **single-cell protein** or **SCP**. SCP has not been universally accepted as a substitute for other types of protein in the human diet; it can be contaminated with other substances and it would have to be made more palatable, but it has been incorporated into animal feeds.

WATER PURIFICATION AND SEWAGE TREATMENT

Many syllabuses require a knowledge of the use of bacteria in the purification of water and in the treatment of sewage.

Water in nature is almost always contaminated with micro-organisms. This is particularly so if faecal matter pollutes the water, when *Vibrio cholerae* and *Salmonella* species (which cause Typhoid) may contaminate the water and be a risk to human health. The common but relatively harmless intestinal bacterium *Escherichia coli*, if present in large numbers in water, indicates contamination with faeces. It is important that water for domestic use and human consumption be purified.

If water is allowed to stand in a reservoir for two or three weeks, there is sedimentation of particles, and as the nutrient level declines, so does the population of micro-organisms. If the water is filtered through fine gravel or sand beds, the bulk of contaminants are removed, and the bacteria, fungi and protozoa which grow in the slime layer over the filter beds decay the organic matter. The addition of chlorine at a concentration of 0.02–2 parts per million will kill any remaining water-borne disease organisms.

River water requires additional processing before it is suitable for human intake. It is coarse-filtered to remove floating material, aerated to reduce odour and treated to reduce mineral content.

GENETIC ENGINEERING

The principles of genetic engineering and some outline of its potential applications should be understood (see Chapter 17). Some syllabuses specify that students should know the details of the production of human insulin and growth hormone using bacteria.

PATHOGENIC ORGANISMS AND DISEASE

Another aspect of microbiology is the effect of the different pathogenic organisms and their control. Bacteria, viruses and fungi can be pathogens of humans, their crops and their domestic animals, and the consequences can be of great economic significance. Some syllabuses include references to the nature of the pathogenic organism, their effects on the hosts and ways in which they can be controlled. Reference is also made to the natural defences of humans, the formation of antibodies and antitoxins.

Disease transmission

Most of the important diseases in man are caused by pathogenic bacteria or viruses. These may be transmitted from one person to another through the air, via contaminated food and water, by direct contact or with the aid of vectors.

Air-borne transmission is also called droplet infection, and is the main way in which diseases of the lungs and respiratory passages are spread. The micro-organisms are expelled in tiny droplets of saliva and mucus, as a result of sneezing, coughing or even just breathing. Bacterial diseases spread this way include whooping cough and tuberculosis, while viral infections include influenza and measles.

Food and water can become contaminated with bacteria if sanitation and hygiene are poor. Salmonella food poisoning is spread in or from the undercrooked meat of infected animals, while diseases such as cholera and typhoid are mainly spread through the contamination of water by the untreated faeces of infected individuals.

Diseases spread by **direct contact** are described as **contagious**. Bacterial infections in this category include the sexually transmitted diseases gonorrhoea and syphilis, and contagious viruses include the cold sore virus, herpes simplex and the AIDS virus, HIV.

Vectors are organisms that carry pathogens from one organism to another and for human diseases these are mainly blood-sucking insects. Typhus is caused by bacteria spread by lice, and yellow fever is caused by viruses carried by *Aedes* mosquitoes. *Anopheles* mosquitoes act as vectors for the protozoan *Plasmodium*, which causes malaria.

Antigens and antibodies

An **antigen** is any substance that when introduced into the blood or tissues induces the formation of antibodies, or reacts with them if they are already present.

An **antibody** is a substance produced by lymphocytes, in the presence of a specific antigen, that can combine with that antigen to neutralise, inhibit or destroy it. Antibodies belong to a group of proteins called globulins (immunoglobulins) and five different classes exist in humans: IgG, IgA, IgM, IgD and IgE. Each has a distinct chemical structure and a specific biological role. IgG antibodies enhance phagocytosis, neutralise toxins (i.e. are antitoxins) and particularly protect the fetus and newborn. IgA antibodies occur on mucous membranes providing local protection. IgM causes agglutination and lysis of microbes. IgD antibodies stimulate antibody-producing cells to produce more antibodies. IgE antibodies are involved in allergic responses.

Various other chemical systems in the body support the action of antibodies but are not themselves antibodies, since they lack specificity. For example, complement is a system of eleven proteins in the plasma, which include **opsonin**. This coats bacteria enabling the phagocytes to engulf them more readily. Cells infected with virus produce a protein called **interferon**, which interferes with the viral replication.

IMMUNITY

Immunity involves two initial processes: the recognition of foreign material when it enters the body and the mobilisation of cells and antibodies capable of removing the foreign material, or antigens, quickly and effectively. A further process involves the formation of memory cells, which carry out the secondary immune response.

There are two systems of immunity in mammals: a cell-mediated immune response and a humoral immune response. Two types of lymphocytes are involved, both of which have receptor sites on their membranes for recognising antigens. Both types are formed in the bone marrow and then complete their development in either the thymus gland, in which case they are known as T-lymphocytes or T-cells, or in lymphoid tissue, producing B-lymphocytes or B-cells.

Cell-mediated Immune Response

The thymus gland is active from birth until a mammal is weaned. During this time it causes lymphocytes to mature and become 'immunologically competent', and capable of synthesising new receptor molecules and incorporating them into the plasma membrane. T-lymphocytes leave the thymus and circulate in the blood and body fluids. If a T-cell meets an antigen for which it has a receptor site, it is stimulated to divide many times by mitosis, forming a clone. Each cell in the clone can attach to a complementary antigen and destroy it.

A further function of T-cells is that their presence is required for the maturation of specific B-lymphocytes.

Humoral Immune Response

B-lymphocytes complete their development in the lymph nodes, or spleen. There are many thousands of different B-lymphocytes, each with just one type of receptor on its cell surface. If an antigen is recognised, the B-cells are stimulated to divide by mitosis forming a clone of plasma cells in the lymph node, and also forming memory cells. The plasma cells live only a few days but can synthesise and secrete vast quantities of specific antibody molecules – nearly 2,000 per second. (These antibodies are of the IgM type.)

The memory cells persist for a long time, sometimes for life. They are responsible for the secondary immune response and confer active immunity against the specific antigen. If that antigen is encountered again, the memory cells can recognise it and stimulate the immediate production of massive quantities of antibody, some IgM as for the primary immune response and also large amounts of IgG.

Types of immunity

There are three types of immunity.

- **Hereditary immunity**   This occurs passively due to the inheritance of genes for disease resistance.
- **Naturally acquired immunity**   This may occur passively or actively.
  - Natural passive immunity may be due to the transfer of antibodies from mother to fetus across the placenta, or from mother to new-born offspring via colostrum, the first secretions of the mammary gland. The immunity is only temporary, since no memory cells have been formed.
  - Natural active immunity is achieved as a result of exposure to infection. The body manufactures its own antibodies in response to the presence of antigens on the infectious agent and also forms specific memory cells. If the same agent is encountered again, it can be flooded with antibodies and eliminated before it causes disease.
- **Artificially acquired immunity**   This may also occur passively or actively.
  - Artificial passive immunity results from the injection of ready-made antibodies into the body, and again, since there are no memory cells, it is only temporary. It is useful as a preventative measure for diseases that are difficult to immunise against, such as tetanus and diphtheria. It may also be used as a treatment for certain diseases, such as rabies, where infection has already occurred which is too dangerous to leave to the body's natural immune system.
  - Artificial active immunity is achieved by **immunisation** or vaccination, in which **vaccine** is injected into a healthy individual. The body is stimulated to produce antibodies and memory cells against the antigen in the vaccine, and thus acquires immunity to subsequent infection by that disease organism.

PRACTICAL WORK

One syllabus suggests that practical work could include plasmid transfer using *Escherichia coli* and selection in *Saccharomyces*. If such practical work is undertaken, it must be done under supervision and under sterile conditions. There are many suggestions for other types of practical work, too numerous to include here.

Further reading

Williams J I and Shaw M, 1982, *Microorganisms*. Bell and Hyman
Staples D G, 1973, *An Introduction to Microbiology*. Macmillan
Land J B and Land R B, 1983, *Food Chains to Biotechnology*. Nelson
Wiseman A (ed), 1983, *Principles of Biotechnology*. Surrey University Press

CHAPTER 18 EXAMINATION QUESTIONS

EXAMINATION QUESTIONS AND ANSWERS

QUESTIONS

Question 18.1
Briefly describe the following techniques, in each case indicating why the technique is used:
a) pasteurisation *(3)*
b) turbidity measurements *(3)*

(Total 6 marks)
(ULEAC AS paper)

Question 18.2
Read the following passage and use the information to answer the questions that follow.

The soil fungus *Pythium* causes a disease known as 'damping off'. The mycelium of the fungus can feed saprophytically on organic matter in the soil, and can infect germinating seeds or young seedlings. Its actively growing hyphae spread through the soil very quickly and it reproduces asexually to form motile zoospores. The fungus usually infects seedlings at, or slightly below, the soil surface. Hyphae penetrate the epidermis of the host and then spread through the cortex by growing between the cells. *Pythium* secretes pectinolytic enzymes which separate the cells thus weakening the seedling, which eventually falls over and dies. Young rootlets of mature plants are also attacked by *Pythium*, but this is rarely harmful to the host. In addition, hyphae cannot penetrate the epidermis of well established plants.

Control methods in the greenhouse involve soil sterilisation and avoiding overcrowding, overwatering and poor ventilation. Fungicides are also used as seed dressings: metalaxyl was introduced in 1977–78, but was soon withdrawn as resistance to this fungicide had developed.

To compare the effectiveness of methods for controlling *Pythium*, an experiment was performed with pea seeds. Seven trays containing sterile compost were each sown with fifty pea seeds. As indicated in the following table, four of the trays, labelled A to D, were inoculated with 50 g of a sand-oatmeal culture of *Pythium* which had been steam treated for 0, 2, 4 and 6 minutes. The seeds in tray E were dressed with the fungicide captan, and the seeds in tray F with the fungicide ethirimol. These two trays were also inoculated with the *Pythium* culture. Tray G contained no inoculum.

After three weeks in a well ventilated greenhouse, the trays were scored for germination of the seeds and the growth of the seedlings. The results are shown in the table below.

TRAY	INOCULUM OF PYTHIUM	TREATMENT Steam sterilisation /min	Fungicide	NUMBER OF SEEDS GERMINATED	GROWTH OF SEEDLINGS
A	✓	0	—	0	–
B	✓	2	—	25	++
C	✓	4	—	47	+++
D	✓	6	—	46	+++
E	✓	—	captan	38	++
F	✓	—	ethirimol	4	+
G	✗	—	—	47	+++

Key: + very limited growth; ++poor growth; +++good growth.

a) Describe the effect of the steam treatment on the sand-oatmeal inoculum of *Pythium*, as revealed by the results given in the Table (trays A to D) *(4)*
b) Comment on the effects of the fungicide dressing, captan and ethirimol. *(4)*
c) Explain why trays A and G were included in the experiment. *(4)*
d) Suggest why oatmeal was used to make the inoculum. *(1)*
e) Give reasons for the poor growth of the seedlings in trays B and E. *(3)*
f) State **three** reasons why *Pythium* can be a serious infection of seedlings grown in greenhouses. *(3)*

g) The sterilisation of the soil in greenhouses, by forcing steam 20–30 cm beneath the surface for at least 10 minutes is not now a common method. Give **two** reasons for this. *(2)*
h) State **one** problem with the use of fungicide seed dressings. *(2)*
i) Explain why *Pythium* does not cause a disease in mature plants. *(2)*

(Total 21 marks)
(UCLES)

Question 18.3

Describe how you would prepare agar plates for the growth of microorganisms and how you would inoculate a plate with a culture of *Escherichia coli*.

(Total 6 marks)
(ULEAC)

Question 18.4

Bacteria of a single species were exposed to ultra violet (UV) radiation. Five different pure colonies, A to E, were then isolated and grown separately on complete medium. The growth of these colonies was then investigated on different media. Organisms from colonies were transferred from one agar plate to the next using sterile card templates (so that the colonies remained in the same positions relative to each other). In all instances the colonies were grown for 24 hours at 25°C.

The results of these investigations are shown in the diagram. It may help you to know that:

1. minimal medium contains glucose and mineral salts dissolved in agar;
2. penicillin kills only *susceptible* bacteria which are *actively growing*;
3. arginine (Arg), tryptophan (Trp) and tyrosine (Tyr) are amino acids.

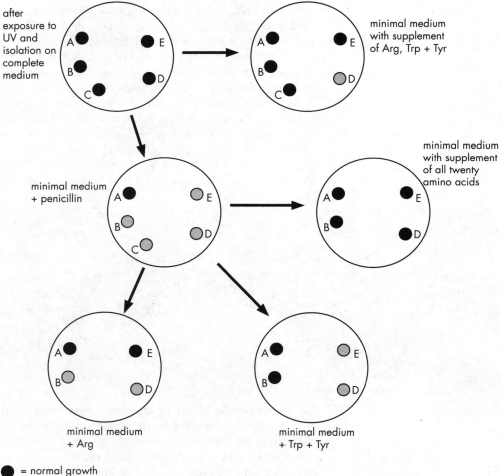

Fig. 18.1
● = normal growth
◯ = slight growth or cells present

Study the diagram of the results carefully and then answer the following questions.
a) Which **one** of the five bacterial colonies, A to E,
  i) is resistant to penicillin?
  ii) requires a supplement of arginine to grow normally?
  iii) appears to have been killed by penicillin?
  iv) requires a supplement other than arginine, tryptophan or tyrosine to grow normally? *(4)*
b) i) What can you deduce about the growth requirements of Colony D? (Be as precise as you can.)
  ii) It can be concluded that Colony C required no supplement to grow on minimal medium. What evidence is there for this conclusion?. *(3)*
c) Design a simple experiment which would enable you to determine more precisely the supplement that Colony B requires for growth. Explain what the results of such an experiment might indicate. *(2)*

(Total 9 marks)
(NEAB)

Question 18.5

The growth of a bacterial culture in the absence of any antibiotic is shown in Fig. 18.2.

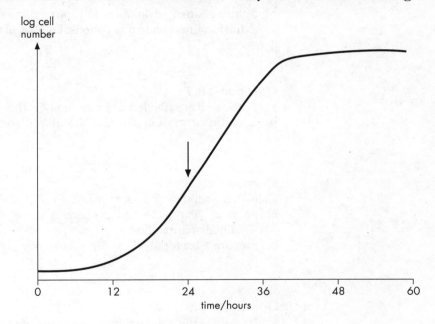

Fig. 18.2

a) Explain what is happening in this culture between:
  i) 0 and 12 hours
  ii) 36 and 48 hours *(4)*
b) Two other cultures of the same bacterium were incubated for 24 hours. After this time, a bacteriostatic antibiotic was added to one culture and a bactericidal antibiotic to the other. On Fig. 18.2, draw clearly labelled lines to show the effects of the two types of antibiotic. The point of addition is shown by the arrow. *(2)*

(Total 6 marks)
(NEAB)

Question 18.6

a) The figure below shows a bacterial suspension on a haemocytometer slide. Assume that the sample shown is representative. In each small square $1/4000$ mm^3 of liquid is trapped beneath the coverslip. Calculate the concentration of bacteria in the suspension. Show your working. *(3)*

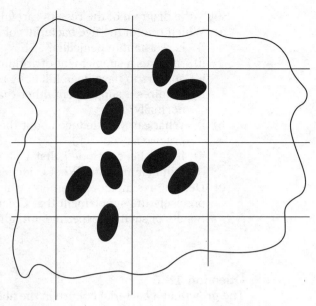

Fig. 18.3

b) Suggest **one** advantage and **one** disadvantage of using this technique for estimating microbial populations, compared with plate counts (viable counts). *(3)*

(Total 6 marks)
(ODLE AS)

Question 18.7
a) Make a large, labelled diagram to show the structure of a named rhizopod. *(6)*
b) State **three** ways in which ciliates differ from rhizopods. *(3)*

(Total 9 marks)
(ULEAC AS paper)

Question 18.8
Using annotated diagrams, show how you would:
a) pour an agar plate and prepare it for storage (starting with a tube of sterile, solidified agar medium)
b) prepare a bacterial smear for microscopy (starting with a broth culture) *(4)*

(Total 4 marks)
(NEAB AS)

Question 18.9
a) How does the animal body react to infection? *(12)*
b) State the principal types of chemotherapeutic agents and briefly indicate the basis of their action in the treatment of infection. *(8)*

(Total 20 marks)
(NEAB AS)

OUTLINE ANSWERS

Answer 18.1
a) Pasteurisation is used in the preservation of milk and beer. It involves heating the liquid at either 63°C for 30 minutes, or 80°C for 30 seconds. The liquid must then be cooled. Most of the contaminating bacteria are then killed.
b) Turbidity measurements are made to measure growth of micro-organisms in culture. A colorimeter is used. It has to be calibrated with a blank tube of medium, then a tube containing the organisms is inserted and the reading recorded.

Answer 18.2
a) Tray A showed that without treatment the fungus will kill all the seedlings. Tray B showed that steam treatment kills the fungus but 2 minutes is too short a time to kill all of it. Only half the seeds germinated and the growth of the seedlings was poor. Trays C and D suggest that 4 minutes of steam treatment killed all the fungal

spores as germination was nearly 100% and seedling growth was good. Tray D showed that steam treatment for more than 4 minutes gave no advantage.
b) Captan reduced the effect of the fungus but only partially as germination was reduced and seedling growth poor. Ethirimol had virtually no effect. Few seeds germinated and seedling growth was poor.
c) Trays A and G were controls. Tray A showed that with no treatment the fungus killed all the seedlings. Tray G showed that the reduction in germination and effects on seedling growth were due to the presence of the fungus. The results in Tray G was the same as that when all the inoculum had been killed.
d) It provided organic matter on which the fungus could feed saprophytically.
e) The fungus infected the plants. Fungal enzymes damaged the cell walls, separating the cells and so disrupting the tissues of the stem cortex.
f) The hyphae grow quickly through the soil, motile zoospores increase the spread of infection, the fungus can feed saprophytically and so survive in the absence of young seedlings.
g) The method is inefficient as a high enough temperature to kill the fungus may not be achieved. The solid will become very wet, increasing the chance of a new infection.
h) Fungicides are toxic to seed eating birds.
i) *Pythium* cannot penetrate the epidermis of mature plants and although it may attack young roots, this does not greatly affect the plant.

Answer 18.3
Use sterile petri dishes. Melt sterile agar and leave to cool until hand-hot. Move the dish lid just sufficient to be able to pour the agar into the dish, pour the agar, replace the lid then leave the dish undisturbed to cool. Flame a wire loop, flame the neck of the culture bottle, dip the loop into the inoculum and replace the lid on the bottle. Do not put the lid down on the bench. With the plate upside down, spread the inoculum onto the plate, turning the plate to spread the bacteria evenly over the whole surface.

Answer 18.4
a) i) A ii) E iii) C iv) D
b) i) It needs at least one amino acid added other than arginine, tryptophan or tyrosine.
   ii) Actively growing cells are killed by penicillin, as shown by the disappearance of C after penicillin treatment. Thus C must grow normally on minimal medium.
c) Set up plates of minimal medium plus tryptophan and minimal medium plus tyrosine. Sub-culture B onto each of the plates. C would only grow actively on the plates containing the necessary supplement.

Answer 18.5
a) i) 0 and 12 hours
      This is the lag phase of growth. The bacteria are adapting to their new environment and have not reached their maximum growth rate.
   ii) 36 and 48 hours
      This is the stationary phase. Growth rate is nearly zero as there is maximum competition for resources and reproduction has almost ceased.
b) Bacteriostatic line would be a plateau at our just above the entry point.
   Bactericidal line would fall steeply to zero at the entry point.

Answer 18.6
a) 8 bacteria in 4 squares;
   so $8 \times 4000/4$ mm^{-3};
   so $8 \times 4000/4 \times 1000$ cm^{-3};
   = 8×10^6 cm^{-3}
b) It is fast, but it does not distinguish between live and dead cells.

Answer 18.7
a) Marks awarded for drawing quality.
   Labels should include: protoplasmic membrane, nucleus, plasmosol, plasmagel, contractile vacuole, food vacuole.

b) Ciliates have a fixed shape, large numbers of cilia, do not form pseudopodia and food vacuoles formed at a special site, the cytostome.

Answer 18.8
a) Agar must be melted, petri dishes must be sterile, agar when liquid must be poured quickly, allowed to solidify and then the plates are inverted to keep the agar dry while stored.
b) Using sterile pipette remove drop of culture and put on slide, smear it over slide and fix it by holding above a flame or drying it in air. Drop stain on the smear, leave for the specified time, wash off, dry and view.

Answer 18.9
a) This part of the answer should deal with the immune response; the ability of the body to recognise foreign material and then the production of substances that will destroy it. A description of the antigen-antibody reaction is needed, with details of the different types of antibodies produced and their functions. It might also be appropriate to elaborate on the types of immunity that there are.
b) Chemotherapeutic agents are chemicals which prevent and cure diseases. Plant extracts were used initially, but now chemicals are produced synthetically. The answer would need reference to sulphonamides and antibiotics, which interfere with bacterial metabolism and so affect their growth. Much of the action of these agents depends on the fact that bacteria are prokaryotic whereas the cells of the animal are eukaryotic and remain unaffected.

QUESTION, STUDENT ANSWER AND EXAMINER COMMENTS

Question 18.10
The diagram below shows the structure of a zygomycete fungus.

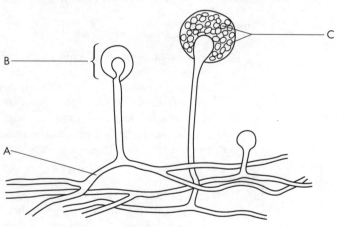

Fig. 18.4

a) i) Identify the fungus.

   Mucor   66 Correct 99   (1)

   ii) Identify the structures labelled A, B and C on the diagram.

   A mycelium

   66 Incorrect, this is a hypha, for the mycelium a bracket label would be used 99

   B sporangium   66 Correct 99

   C spores   66 Correct 99   (3)

b) Describe how this organism survives adverse conditions.

   In adverse conditions mucor forms zygospores which are resistant to cold and drought.

   66 Correct but lacks sufficient detail. Should also be comment on fusion of nuclei or gametangium formation 99

   (2)
   (Total 6 marks)
   (ULEAC)

REVIEW SHEET

1. What do you understand by **fermentation**?

2. Identify some of the main **products of fermentation**.

3. Identify the four stages in **gene manipulation**.
   a)
   b)
   c)
   d)

4. a) Draw a diagram to show the insertion of DNA into a plasmid vector.

   b) Use your diagram to explain the process of **splicing**.

5 a) Define, and explain the uses of, **monoclonal antibodies**.

_____
_____
_____
_____

b) Present a diagram to explain the *technique* of producing monoclonal antibodies.

6 Define what is meant by the term **enzyme** and consider some of their uses.

_____
_____
_____
_____
_____

7 What is meant by **autotrophic micro-organisms**? Give examples and consider their usefulness.

_____
_____
_____
_____
_____

8 What is meant by **heterotrophic micro-organisms**? Again give examples and consider their usefulness.

_____
_____
_____
_____

9 Consider some human applications of DNA technology.

_____
_____
_____
_____
_____

INDEX

abiotic features 230
abortion 158
abscisic acid 105
absorption 60–1
absorption spectrum 43, 44
acetylcholine 120
acrosome 157
action potential 119
action spectrum 44
active transport 134
ADH 72, 73
adaptive radiation 216
adrenal gland 122–3
adrenalin 122
aerobic respiration 86–8
aldosterone 123
allantois 159
alleles 205
allele frequency 213
allopolyploidy 210
amino acids 29
amnion 159
amoeboid movement 190–1
Amphibia 17
anaerobic respiration 88–9
androecium 168
aneuploidy 213
Angiospermaphyta 15, 167
Animalia 15
Annelida 16
Anthozoa 15
antibodies 267
antidiuretic hormone 72, 73
Arachnida 17
Arthropoda 16, 189
artificial selection 217
Ascomycota 14
ATP 86, 87, 88
atrioventricular node (AVN) 138
autonomic nervous system 75, 117–8
autopolyploidy 209
autosomes 211
auxins 102, 104
Aves 17

back cross 211
bacterial nutrition 264–5
balanced diet 58
Basidiomycota 14
bile 60
binomial system 13
biogeochemical cycles 231–3
biogeochemical oxygen demand (BOD) 237
biosensors 253
blastula 158
blastocyst 158
blood 135–6
blood glucose level 101
blood glucose regulation 61
blood vessels 113
Bohr effect 92
bone 186–7
bony fish 76
Bowman's capsule 70–2
brain 114–6
brown fat 74
Bryophyta 14

C_3 and C_4 plants 49
Calvin cycle 47
cambium 174–5
    cork 175
    vascular 174
carbohydrates 48, 89
carbon cycle 232, 233
carbon dioxide (for photosynthesis) 42

cardiac muscle 137
carotenoids 45
carpel 168
cartilage 187–8
cation pump 119
cellulose 28
cerebellum 115
cerebrum 115
Cestoda 16
chiasmata 204
Chilopoda 16
chitin 189
chloride secretory cells 76
Chlorophyta 13
chloroplast 44
chlorophyll 45
cholinesterase 120
Chondrichthyes 17
Chordata 17
chorion 159
chromatids 202
chromosomes 202, 204, 209
chromosome mapping 211
chromosome mutations 213
cilia 191
Ciliphora 14
circulation 135
cistron 32, 205
class 12
climax community 236
Cnidaria 15
collenchyma 192
colostrum 161
community 230
comparative anatomy 216
comparative physiology 180
conditioning 125
Coniferophyta 15
continuous variation 202
contractile vacuole 75
contraception 147–8
corpus luteum 156
cortisol 123
Craniata 17
cranium 184
cretinism 122
cross-over value (COV) 211
Crustacea 16
cytochromes 46
cytokinins 105

deletion 213
denitrification 232
density dependence 236
diabetes mellitus 122
diet 58
diffusion 152
digestion 59–60
dihybrid ratio 207
disaccharide 28
discontinuous variation 202
dorsal root 116
DNA 30, 31–2, 205, 212
DNA fingerprinting 255
DNA ligase 251
dominant allele 206
ductus arteriosus 160
duodenum 59

earthworms 90, 114, 190
ecdysone 124
Echinodermata 17
ecological niche 230
ectoderm 74
edaphic features 230
endoderm 12

endometrium 155, 156
endoplasmic reticulum 33
endotherms 74
enzymes 34–5
    active site 34
    classification 35
    cofactor 24
    inhibition 35
epicuticle 76
epidermis 154
erythrocyte 136
essays 6–7
ethene 105
eutrophication 237
evolution (evidence for) 216
examination strategy 6
excitatory synapses 120
exoskeleton, insects 189
extinction 218

fallopian tube 155
facilitated diffusion 132
FAD 87
families 12
fat 89
feeding mechanisms 57
fertilisation 172
fetal circulation 160–1
field work 9
Filicinophyta 15
flagella 191
flowers 168–172
fluid feeders 57
fluid mosaic model 33
follicle stimulating hormone (FSH) 156
fossil dating 216
fossil evidence 216
fountain zone theory 191
Fungi 14, 264, 265

gametogenesis 155
gas exchange 90–4
    animals 90–2
    plants 93–4
Gastropoda 16
gene 205
    interaction 208
    manipulation 250, 251
    mutation 213
    pool 213, 217
genetic information transfer 251
genotype 206
genus 12
geographical distribution 216
geotropism 103, 104, 105
germination 104, 172, 173
gibberellins 104–5
glucagon 73, 122
glucose 27
glycolysis 86
glycosidic bond 28
Graafian follicle 156
gynaecium 168

habitat 230
habituation 125
Hardy-Weinberg equation 214
haversian systems 187
head 137, 138
Hepaticase 14
heterosomes 211
heterozygous 206
hindbrain 115
holophytic nutrition 42
holozoic nutrition 56
homologous chromosomes 204

INDEX

homozygous 206
hormones 75
hybridomas 253
Hydra 114
Hydrozoa 14
hyperthyroidism 122
hypothalamus 75, 122

ileum 59
implantation 158
immunity 267–8
imprinting 124
incomplete dominance 208
infertility 158
inhibitory synapses 120
innate behaviour 124
Insecta 16
insects
    flight in 191
    hormones 123–4
    walking in 191
    water balance 76
insulin 73, 122, 252
inversion 213
ionising radiation 212
islets of Langerhans 122
isolation 217–8

joints 185, 186, 187
juvenile hormone 123

key words 6
keys 13
kidney, mammalian 70–3
kinesis 124
kingdom 12
Krebs cycle 86, 87, 88

lactation 161
leaf structure 45
learned behaviour 124
lenticels 93
leucocyte 126
Leydig cells 154
light (in photosynthesis) 43
lignin 173
limb girdles 185
limiting factors 48
linkage groups 210
lipids 29
liver 61–2, 73–4
locus 205
loop of Henle 70–2
luteinising hormone 156
Lycopodophyta 15
lymph 138

macrophagous feeders 57
malpighian tubules 76
Mammalia 17
meiosis 204–5
Mendelian laws 206–9
microhabitats 230
meristems 173, 174
mesoderm 12, 159
mesogloea 114
microbial enzymes 251
microbial mass 251
microbial metabolites 251
microfibrils 18
microphagous feeders 57
microscopy 9
mitochondria 89
mitosis 202–3
Mollusca 16
monamino oxidases 120
monoclonal antibodies 253, 254
monohybrid inheritance 206
monosaccharides 27
multicellular organisms 12
multiple alleles 208
multiple choice 7–8

Musci 14
mutagenic agents 212
mutons 205

NAD 87
NADP 45, 46, 47
nastic movements 106
natural selection 215
negative feedback 122
Nematoda 16
neo-Darwinism 215
nephron 70–2
nerve impulses 118–19
neurones 118
neurotransmitters 118, 119
Nissl granules 118
nitrification 232
nitrogen cycle 232
nitrogen fixation 232
node of Ranvier 118
non-disjunction 213
noradrenalin 120, 122
nucleic acids 30
nucleotides 30

oestrogen 156
Oligochaeta 16
Oomycota 13
orders 12
ornithine cycle 62, 74
Osteichthyes 17
osmosis 132-134
ova 156
ovaries 154, 169
ovules 169
oxidative phosphorylation 86
oxygen carriage 92
oxygen debt 189
oxytocin 157

pacemaker 173
pancreas 122
pancreatic juice 60
parasite 56
parasitic nutrition 56
parasympathetic system 157
parenchyma cells 192
parental care 161
pectoral girdle 185
pelvic girdle 185
Pelycipoda 15
peripheral nervous system 116
Phaeophyta 14
phages 251
phagocytosis 134
phenotype 206
phloem 141–142
phosphoglyceric acid (PGA) 47
phospholipids 29
phosphorus cycle 233
photorespiration 49
photosynthetic pigments 44
photosystems 46, 47
phototropism 102–3, 105
phyla 12
phytochrome 106
pinocytosis 134
pioneer community 236
pituitary gland 122
placenta 160
plagioclimax 236
Plantae 14
plasma 135
plasmids 251
Platyhelminthes 16
polar molecule 26
pollen grains 168
pollination 170, 171
Polychaeta 16
polygenes 208
polypeptide 29
polysaccharide 28

population 30
Porifera 15
practical papers 8
progesterone 156
projects 9
Prokaryotae 13, 264
prolactin 156
propagules 175
prostaglandins 157, 161
proteins 29–30, 48, 89
Protoctista 13, 264
protozoa 75
Punnett square 207
purine 30
Purkinje tissue 138
putrefaction 231
pyrimidines 30
pyramids, of biomass 235
    of energy 235
    of numbers 234

recessive alleles 206
reflex arc 116
refractory period 119
Reptilia 17
resource management 237
respiratory quotient 89
    substrates 89
    surfaces 90
resting potential 118
restriction enzymes 251, 255
reverse transcriptase 251
Rhizopoda 13
ribosome 32
ribulose bisphosphate 47
RNA 30–2
    messenger 32
    transfer 32
ruminant 60

saprotrophic nutrition 56–7
Schwann cells 118
sclerenchyma 192
secondary xylem 175
seed development 172
selective breeding 217
selective reabsorption 71–2
Sewall Wright effect 214
short answer papers 6
skull 184
simple reflex 124
single cell protein 266
sino-atrial node 138
social behaviour 125
species 12
Spenophyta 15
spermatozoa 154, 155
spinal cord 116
splicing (of DNA fragments) 251
stamens 168
starch 28, 47
sterilisation 158
stigma 168
stomach 59
stomata 92
striated muscle 188–9
subspecies 12
sulphur cycle 232
sulphur diozide 236
surface tension 27
sutures 184
sympathetic system 117, 118
synapses 119–120

T system 188
tactic movements 106
taxis 124
teacher assessment 8–9
teeth 58
teleosts 76
testes 154
testosterone 154

thrombocyte 136
thyroid gland 122
thyroxine 75, 122
tracheids 139, 142, 192
transcription 32
transduction 121
translation 32
translocation 213
transpiration 140
Trematoda 16
triglyceride 29
triploblastic 12
trophic level 233
tropic movements 105, 106

tropomysin 188
troponin 188
Turbellaria 16

ultrafiltration 71
umbilical artery and vein 160
unicellular 12
urea 74

varieties 12
vascular tissues 139
ventral root 116
vertebral column 184
Vertebrata 17

vessels 139, 192
vestigial organs 216
villus 60–1
viruses 264
vitamins 58, 61

water 23–7
   in photosynthesis 43
   potential 133

xanthophylls 45
xylem 139, 192–3

zwitterions 29

280

ORDER FORM

If you have enjoyed using this book and would like to purchase or see inspection copies (Booksellers and Teachers/Lecturer only) of any of the other books available in the Longman GCSE, A-level or GNVQ series, then follows these simple procedures

- Select the tiles you require from the list below.
- Complete the order from over-leaf.
- Complete the appropriate delivery instructions.
- Select your method of payment.
- Detached this page and send or fax it to the following address.

  Longman Group Ltd.
  PO Box 88
  Harlow
  Essex CM19 5SR
  Fax no. (01279) 453450

- Or make your order by telephone (01279) 623923 quoting the ISBN, title and quantity required.

AVAILABLE TITLES

LONGMAN GCSE/KS4 REVISE GUIDES

ISBN	TITLE	PRICE
23773 6	Biology N/E	£9.99
27634 2	Business Studies N/E	£9.99
23773 4	Chemistry N/E	£9.99
24685 3	Economics N/E	£9.99
22832 8	English N/E	£9.99
22831 X	English Literature N/E	£9.99
23775 0	Higher Level Mathematics N/E	£9.99
24494 3	Information Systems	£10.99
22830 1	Mathematics N/E	£9.99
23774 2	Physics N/E	£9.99
27687 X	Religious Studies N/E	£9.99
22829 8	Science N/E	£9.99
23771 8	Technology N/E	£9.99

LONGMAN GCSE REVISE GUIDES

ISBN	TITLE	PRICE
01884 6	Art and Design	£8.99
03854 5	Computer Studies	£8.99
02506 0	French	£8.99
03839 1	French pack	£12.99
03836 7	French cassette	£5.00
02505 2	Geography	£8.99
02430 7	German	£8.99
03837 5	German pack	£12.99
03838 3	German cassette	£5.00
03855 3	Home Economics	£8.99
22845 X	Music N/E	£9.99
22651 1	Sociology	£8.99
22652 X	Spanish	£8.99
24509 5	Spanish pack	£13.99
24511 7	Spanish cassette	£6.00
01578 2	World History	£8.99

LONGMAN A-LEVEL REVISE GUIDES

ISBN	TITLE	PRICE
22569 8	Accounting	£10.99
05780 9	Art and Design	£9.99
27771 X	Biology N/E	£9.99
08909 3	Business Studies	£9.99
05179 7	Chemistry	£9.99
05782 5	Computer Science	£9.99
27688 8	Economics N/E	£10.99
05175 4	English	£9.99
05784 1	French	£9.99
24495 1	French pack	£14.99
24497 8	French cassette	£6.00
22570 1	General Studies	£9.99
05173 8	Geography	£9.99
22653 8	German	£9.99
24498 6	German pack	£14.99
24508 7	German cassette	£6.00
27689 6	Mathematics N/E	£10.99
05177 0	Modern History	£9.99
27690 X	Physics N/E	£9.99
22654 6	Psychology	£9.99
05786 8	Sociology	£9.99

LONGMAN ADVANCED GNVQ TEST AND ASSESSMENT GUIDES

ISBN	TITLE	PRICE
23776 9	Business	£12.99
24634 2	Construction and the Built Environment	£14.99
23777 7	Health and Social Care	£12.99
24635 0	Hospitality and Catering	£14.99
23778 5	Leisure and Tourism	£12.99
24633 4	Science	£14.99

Please note that prices and other details are correct at time of going to press, but are liable to subsequent change without notice.

ISBN	TITLE	QUANTITY	TOTAL PRICE

Students only. *Please add postage and packaging as follows: orders up to £9.99, £0.78; £10.00–£14.99, £1.20; £15.00–£19.99, £1.44; £20.00–£29.99, £2.42; £30.00–£49.99, £3.62; £50.00 and over, £4.85.*

TOTAL POSTAGE _____

TOTAL COST _____

DELIVERY INSTRUCTIONS

Complete the appropriate section below, clearly stating address and special delivery instructions.

BOOKSELLERS

Address _____ Order Reference _____

_____ Special instructions for delivery _____

_____

_____

STUDENTS

Name _____

Address _____

_____

Post code _____

TEACHERS AND LECTURERS

Teachers and Lecturers who are considering recommending Longman Revise Guides to their students may order up to three copies on 28 days' FREE inspection. Please list the titles below.

_____

_____

_____

Place an order for 12 copies or more per title for your school or college and receive an educational discount. Simply complete and return the order form. This offer applies to orders supplied direct from Longman only.

Mr/Mrs/Miss/Ms _____

Department _____

School/College _____

Address _____

_____

Post code _____

Longman Customer Account No. *(if known)* _____

METHOD OF PAYMENT

Please tick the relevant box and complete.

❏ I enclose a cheque for P/O for £_____ made payable to Longman Group Ltd.

❏ Please charge my Visa/Access/AMEX/Diners Club card.

Number ☐☐☐☐ ☐☐☐☐ ☐☐☐☐ ☐☐☐☐

Expiry date _____

Signature _____

Charge address on card, if different to the address above _____

❏ Please invoice me.